Lecture Notes in Mathematics

A collection of informal reports and seminars
Edited by A. Dold, Heidelberg and B. Eckmann, Zürich

162

Harish-Chandra
The Institute for Advanced Study
School of Mathematics, Princeton, NJ/USA

Notes by

G. van Dijk

Harmonic Analysis on Reductive p-adic Groups

Springer-Verlag
Berlin · Heidelberg · New York 1970

 Library of Congress Catalog Card Number 79-138810 Printed in Germany.
Title No. 3319

Offsetdruck: Julius Beltz, Weinheim/Bergstr.

CONTENTS

Part I. Existence of characters for the discrete series.

§1. Square-integrable representations mod Z. 4
§2. Reductive $\mathfrak{p}$-adic groups. 8
§3. Supercuspidal representations. 9
§4. A conjecture. 10

Part II. Existence of characters in the general case.

§1. The Godement principle. 12
§2. A theorem of Bruhat and Tits. 15
§3. Proof of Theorem 4 (based on Conjecture I). 16

Part III. Supercusp forms and supercuspidal representations.

§1. The space generated by a supercusp form. 20
§2. Some consequences. 26

Part IV. The space $\mathcal{A}(G, \tau)$.

§1. Conjecture III. 30
§2. The space $\mathcal{A}(G, \tau)$. 31
§3. Proof of Theorem 7. 38
§4. Proof of Theorem 8. 39

Part V. The behavior of the characters of the supercuspidal representations on the regular set.

§1. Two fundamental theorems. 43
§2. $\mathfrak{p}$-Adic manifolds and distributions. 48
§3. Invariant distributions on the regular set. 51
§4. Applications to the characters of the supercuspidal representations. 57

Part VI. The mapping "F_f" (char $\Omega = 0$).

§1. Introduction and elementary properties of the mapping "Φ_f". 63
§2. The first step in the proof of Theorem 13. 68
§3. Some algebraic lemmas on nilpotent elements. 71
§4. A submersive map. 72
§5. Some more preparation. 76
§6. The second step in the proof of Theorem 13. 78
§7. Completion of the proof of Theorem 13. 81
§8. Lifting of Theorem 13 to the group. 82

Part VII. The local summability of $|D|^{-\frac{1}{2}-\varepsilon}$ (char $\Omega = 0$).

§1. Statement of Theorem 15. Reduction to the Lie algebra. 86
§2. Proof of the main lemma. 90

Part VIII. The local summability of the characters of the supercuspidal representations (char $\Omega = 0$).

§1. The main theorem and its consequences. 92
§2. Statement of the preparatory results for the proof of Theorem 16. 95
§3. Proof of the main theorem. 98
§4. Proof of Lemma 46. 103
§5. Proof of Theorem 18 (first step). 106
§6. Proof of Theorem 19. 108
§7. Proof of Theorem 18 (second step). 114
§8. Proof of Theorem 20. 116

Introduction

The object of these lectures is to illustrate, what I like to call the Lefschetz principle, which, in the context of reductive groups, says that whatever is true for real groups is also true for $\mathfrak{p}$-adic groups. The theory of reductive algebraic groups can be divided into four parts arranged according to their increasing transcendental component.

1) Algebraic theory.

2) Structure theory.

3) Local Fourier Analysis.

4) Global Fourier Analysis on G_A/G_F.

With the work of Chevalley and that of Borel and Tits, the Lefschetz principle may now be regarded as quite well established for the algebraic theory. Moreover the recent work of Bruhat and Tits shows that this principle also holds with regard to the structure theory of real and $\mathfrak{p}$-adic groups. The results of Mautner, Bruhat, Satake, Gelfand-Graev, Macdonald, and Sally-Shalika suggest that it is also valid at stage 3). These lectures are meant to strengthen this conclusion. Finally the work of Jacquet and Langlands on automorphic forms on GL(2), shows that all primes ought to be treated on an equal footing and this is precisely the content of the Lefschetz principle in the global case.

These lectures may also be regarded as an attempt to justify a claim about the philosophy of cusp forms which I made some years ago (see "Eisenstein series over finite fields," Stone Jubilee volume, Springer). We shall see how this philosophy can be used to suggest, and occasionally to prove, results on harmonic analysis on reductive $\mathfrak{p}$-adic groups.

Let Ω be a $\mathfrak{p}$-adic field (i.e. a locally compact field with a discrete valuation) and let $\underset{\sim}{G}$ be a linear algebraic group defined over Ω. We assume that $\underset{\sim}{G}$ is connected and reductive. Let G denote the set of all Ω-rational points in $\underset{\sim}{G}$. Then G is a closed subgroup of $GL(n, \Omega)$ and so it is locally compact. Also it is separable and unimodular.

Let dx denote the Haar measure on G and $C_c^{\infty}(G)$ the space of all complex-valued locally constant functions on G with compact support. Let π be an irreducible unitary representation of G on a Hilbert space $\mathcal{H}$. For any $f \in C_c^{\infty}(G)$, define the operator

$$\pi(f) = \int_G f(x)\pi(x)dx \ .$$

The following two statements are equivalent.

1) Let K be an open compact subgroup of G and $\underline{d}$ an irreducible representation of K. Then the multiplicity of $\underline{d}$ in the restriction of π to K is finite.

2) For any $f \in C_c^{\infty}(G)$, $\pi(f)$ is of the trace class.

By analogy with the real case, one expects them to be true. Assuming that this is so, put

$$\Theta_\pi(f) = \operatorname{tr} \pi(f) \ .$$

Then Θ_π is a distribution on G which depends only on the class ω of π. Hence we may denote it by Θ_ω. It is easy to show that the mapping $\omega \mapsto \Theta_\omega$ is injective. If $\ell = \operatorname{rank} \underset{\sim}{G}$, let $D(x)$ $(x \in G)$ denote the coefficient of t^ℓ in the polynomial $\det(t+1-\operatorname{Ad}(x))$ in the indeterminate t. Let G' be the set of all $x \in G$ where $D(x) \neq 0$. Then G' is an open dense subset of G whose complement is of measure zero. Again by analogy with the real case, one expects that the following statements are true.

3) Θ_ω is a locally summable function on G which is locally constant on G'.

4) The function $\Phi_\omega = |D|^{1/2}\Theta_\omega$ is locally bounded on G.

By a Cartan subgroup A of $\underset{\sim}{G}$, we mean a subgroup of the form $A = \underset{\sim}{A} \cap G$ where $\underset{\sim}{A}$ is a maximal Ω-torus in $\underset{\sim}{G}$. For $f \in C_c^{\infty}(G)$, define

$$F_f^A(a) = F_f(a) = |D(a)|^{1/2} \int_{G/A} f(xax^{-1})dx^* \qquad (a \in A' = A \cap G') \ ,$$

where dx^* is the invariant measure on G/A. It is easy to see that F_f is locally constant on A'. The Lefschetz principle leads us to expect the following.

5) F_f remains bounded on A. Moreover at any point $a_0 \in A$, F_f has only finitely many distinct limits.

I think it is very important to investigate these limits. (Those familiar with the real case, would see the analogy immediately.) The significant case is when A is compact.

Finally we come to the following question.

6) What is the connection, if any, between the functions F_f^A (for the various Cartan subgroups A of G) and $f(1)$?

Of course the main goal here is the Plancherel formula. However, I hope that a correct understanding of this question would lead us in a natural way to the discrete series for G. (This is exactly what happens in the real case. But the $\mathfrak{p}$-adic case seems to be much more difficult here.)

I propose to discuss some of these questions in these lectures and offer some partial answers which seem to be quite encouraging.

These lectures were given at The Institute for Advanced Study during the Fall Term of '69. I am very grateful to Dr. van Dijk for having taken the trouble of preparing these notes. In addition to streamlining the presentation, he has had to work out the details of many proofs and although I have not personally checked them, I trust there are no mistakes.

Harish-Chandra

Part I. Existence of characters for the discrete series.

§1. Square-integrable representations mod Z.

Let G be a locally compact, separable and unimodular group with center Z_G. Denote by Z a closed subgroup of Z_G such that the quotient Z_G/Z is compact. Let $\hat{Z}$ stand for the set of all (unitary) characters of Z. Fix Z and $\chi \in \hat{Z}$.

By π we denote a (continuous) unitary representation of G on a Hilbert space $\mathcal{H}$. We call π a χ-representation if $\pi(z) = \chi(z)\cdot 1$ for all $z \in Z$.

Now let π be an irreducible χ-representation.

Definition. π is said to be square-integrable mod Z if there exist $\phi, \psi \in \mathcal{H} - \{0\}$ such that

$$\int_{G/Z} |(\phi, \pi(x)\psi)|^2 dx^* < +\infty .$$

Let $C_c(G, \chi)$ be the space of continuous complex-valued functions f on G with compact support mod Z (i.e. $\operatorname{Supp} f \subset C_f . Z$, C_f being a compact subset of G, depending on f), satisfying $f(xz) = f(x)\chi(x)$ $(x \in G,\ z \in Z)$. Denote by $L_2(G, \chi)$ its completion with respect to the norm, denoted $\| \ \|$, derived from the scalar product $(f, g) = \int_{G/Z} \overline{f(x)}\, g(x) dx^*$ $(f, g \in C_c(G, \chi))$. Let λ be the left-translation on $L_2(G, \chi)$:

$$\lambda(x)f(t) = f(x^{-1}t) \qquad (x, t \in G) .$$

Then λ is a continuous unitary representation of G on $L_2(G, \chi)$. Actually $\lambda = \operatorname{ind}_{Z\uparrow G} \overline{\chi}$, the unitary representation of G induced by $\overline{\chi}$ in the sense of Mackey. We call λ the left regular representation of G on $L_2(G, \chi)$. The following theorem is standard. It goes back to Mackey (unpublished).

Theorem 1 (Schur orthogonality relations). Let $(\pi, \mathcal{H})$ and $(\pi', \mathcal{H}')$

be irreducible χ-representations of G.

(a) The following conditions are equivalent:

(i) π is square-integrable mod Z,

(ii) $\int_{G/Z} |(\phi, \pi(x)\psi)|^2 dx^* < +\infty$ for every pair $\phi, \psi \in \mathcal{H}$,

(iii) π is equivalent to a subrepresentation of $(\lambda, L_2(G, \bar{\chi}))$.

If these conditions are fulfilled, there exists a number $d(\pi) > 0$, called the formal degree of π, depending only on the normalization of the Haar measure on G/Z, such that

$$\int_{G/Z} \overline{(\phi_1, \pi(x)\psi_1)}\ (\phi_2, \pi(x)\psi_2) dx^* = d(\pi)^{-1}\ \overline{(\phi_1, \phi_2)}\ (\psi_1, \psi_2)$$

for all $\phi_i, \psi_i \in \mathcal{H}$ (i = 1, 2).

(b) If π is not equivalent to π', then

$$\int_{G/Z} \overline{(\phi_1, \pi(x)\psi_1)}\ (\phi_2, \pi'(x)\psi_2) dx^* = 0$$

for all $\phi_i \in \mathcal{H}$, $\psi_i \in \mathcal{H}'$ (i = 1, 2).

Denote by $\mathcal{E}(G)$ the set of equivalence classes of irreducible unitary representations of G and by $\mathcal{E}(G, \chi)$ the subset consisting of those classes, which contain χ-representations. Let $\mathcal{E}_2(G)$ be the subset of $\mathcal{E}(G)$, consisting of the classes which contain square-integrable representations mod Z and put $\mathcal{E}_2(G, \chi) = \mathcal{E}_2(G) \cap \mathcal{E}(G, \chi)$. We call $\mathcal{E}_2(G)$ the discrete series of G. It is independent of the choice of Z.

Observe that $d(\pi)$ only depends on the class ω of π. We shall write $d(\omega)$ as well as $d(\pi)$.

Let $\omega \in \mathcal{E}_2(G, \chi)$ and fix $\pi \in \omega$. Denote by M_π the closed subspace of $L_2(G, \chi)$, generated by the matrix coefficients of π. For obvious reasons we shall write M_ω as well as M_π. Denote by E the orthogonal projection of $L_2(G, \chi)$ on the orthogonal sum of the spaces M_ω $(\omega \in \mathcal{E}_2(G, \chi))$ and let E_ω be the orthogonal projection on M_ω. Clearly $E = \sum_{\omega \in \mathcal{E}_2(G,\chi)} E_\omega$.

Lemma 1. Let $\omega \in \mathcal{E}_2(G, \chi)$ and fix $\pi \in \omega$. For $f \in C_c(G, \chi)$ the operator

$$\pi_*(\bar{f}) = \int_{G/Z} \bar{f}(x)\pi(x)dx^*$$

is of Hilbert-Schmidt type and its H-S norm, denoted by $\|\pi_*(\bar{f})\|_2$ satisfies

$$\|\pi_*(\bar{f})\|_2^{\,2} = d(\omega)\,\|E_\omega f\|^2 .$$

Proof. Let $(e_i)_{1 \le i < \infty}$ be an orthonormal base of $\mathcal{H}$, the representation space of π. Define $c_{\pi, e_i, e_j}(x) = (e_i, \pi(x)e_j)$ $(x \in G;\ 1 \le i, j < \infty)$. The vectors $f_{ij} = d(\omega)^{1/2} c_{\pi, e_i, e_j}$ form an orthonormal base for M_ω. Hence $\|E_\omega f\|^2 = \sum_{i,j} |(f_{ij}, f)|^2 = d(\omega) \sum_{i,j} |(c_{\pi, e_i, e_j}, f)|^2 = d(\omega) \sum_{i,j} |(e_i, \pi_*(\bar{f})e_j)|^2$ $= d(\omega) \|\pi_*(\bar{f})\|_2^{\,2}$. This proves the lemma.

Now let K be a compact subgroup of G. By $\mathcal{E}(K)$ we denote the set of equivalence classes of irreducible unitary representations of K. If $\omega \in \mathcal{E}(G)$ and $\underline{d} \in \mathcal{E}(K)$, we define $[\omega : \underline{d}]$ as follows. Fix a representation $\pi \in \omega$ and let π_K denote the restriction of π to K. Then $[\omega : \underline{d}]$ is the multiplicity of $\underline{d}$ in π_K. Since K is compact, every irreducible representation of K is finite-dimensional. We write $d(\underline{d})$ for the degree of a representation in the class $\underline{d}$. By $\alpha_{\underline{d}}$ we denote the function on K given by $\alpha_{\underline{d}}(k) = d(\underline{d})\ \overline{\mathrm{tr}\ \underline{d}(k)}$. We have the following theorem.

Theorem 2. Let K be an open compact subgroup of G. Normalize the Haar measures dx and dz on G and Z such that the total measures of K and $K \cap Z$ are 1. Normalize the Haar measure dx^* on G/Z such that $dx = dx^* dz$. Then for any $\underline{d} \in \mathcal{E}(K)$ and $\chi \in \hat{Z}$,

$$\sum_{\omega \in \mathcal{E}_2(G, \chi)} d(\omega)[\omega : \underline{d}] \le d(\underline{d}) .$$

Proof. If $f \in C_c(G, \chi)$, then by Lemma 1,

$$\|f\|^2 \geq \|Ef\|^2 = \sum_{\omega \in \mathcal{E}_2(G,\chi)} \|E_\omega f\|^2 = \sum_{\omega \in \mathcal{E}_2(G,\chi)} d(\omega) \|\pi_*(\bar{f})\|_2^2 ,$$

where π is any, but for the time being, fixed, representation in the class ω ($\omega \in \mathcal{E}_2(G, \chi)$).

Choose f as follows. Extend $a_{\underline{d}}$ to G by requiring $a_{\underline{d}}(x) = 0$ if $x \in {}^c K$, and put $\bar{f}(x) = \int_Z a_{\underline{d}}(xz)\chi(z)dz$ $(x \in G)$. Then

$$\pi(a_{\underline{d}}) = \int_G a_{\underline{d}}(x)\pi(x)dx = \int_{G/Z} dx^* \int_Z a_{\underline{d}}(xz)\pi(xz)dz$$

$$= \int_{G/Z} \bar{f}(x)\pi(x)dx^* = \pi_*(\bar{f}) .$$

Therefore $\|\pi_*(\bar{f})\|_2^2 = d(\omega)[\omega : \underline{d}]$. On the other hand

$$\|f\|^2 = \int_{G/Z} |f(x)|^2 dx^* = \int_{K/Z\cap K} |f(k)|^2 dk^* = \int_K |f(k)|^2 dk .$$

But, observing that for any $\tau \in \underline{d}$, $\tau(z) = \eta(z)\cdot 1$ $(z \in K \cap Z)$ for some $\eta \in (K \cap Z)^\wedge$, we obtain

$$\bar{f}(k) = \int_Z a_{\underline{d}}(kz)\chi(z)dz = \int_{Z\cap K} a_{\underline{d}}(k)\eta(z^{-1})\chi(z)dz = \begin{cases} 0 & \text{if } \eta \neq \chi_K \\ a_{\underline{d}}(k) & \text{if } \eta = \chi_K \end{cases}$$

where χ_K is the restriction of χ to $Z \cap K$. In any case

$$\|f\|^2 \leq \|a_{\underline{d}}\|^2 = d(\underline{d})^2 .$$

The assertion of the theorem is now immediate.

Corollary. Fix $\omega \in \mathcal{E}_2(G)$. Under the conditions of Theorem 2,

$$[\omega : \underline{d}] \leq d(\omega)^{-1} d(\underline{d})$$

for all $\underline{d} \in \mathcal{E}(K)$.

§2. Reductive $\mathfrak{p}$-adic groups.

We recall here some standard facts about algebraic groups (cf. A. Borel - J. Tits, Groupes réductifs, Publ. Math. I.H.E.S. 27 (1965), 55-152). Let Ω be a field. By an Ω-group, we mean a (linear) algebraic group defined over Ω. Let $\underset{\sim}{G}$ be a connected and reductive Ω-group. By a parabolic subgroup $\underset{\sim}{P}$ of $\underset{\sim}{G}$ we mean an algebraic subgroup which contains a Borel subgroup of $\underset{\sim}{G}$. We say that $\underset{\sim}{P}$ is Ω-parabolic if it is parabolic and defined over Ω. Let $\underset{\sim}{N}$ denote the unipotent radical of $\underset{\sim}{P}$. Then $\underset{\sim}{N}$ is an Ω-subgroup and $\underset{\sim}{P}$ is the normalizer of $\underset{\sim}{N}$ in $\underset{\sim}{G}$. By a Levi Ω-subgroup $\underset{\sim}{M}$ of $\underset{\sim}{P}$ we mean a reductive Ω-group such that the mapping $(m, n) \longmapsto mn$ $(m \in \underset{\sim}{M},\ n \in \underset{\sim}{N})$ defines an Ω-isomorphism of the algebraic varieties $\underset{\sim}{M} \times \underset{\sim}{N}$ and $\underset{\sim}{P}$. Such a subgroup $\underset{\sim}{M}$ always exists and is connected. Fix $\underset{\sim}{M}$ and let $\underset{\sim}{A}$ be a maximal Ω-split torus lying in the center of $\underset{\sim}{M}$. Then $\underset{\sim}{A}$ is unique and $\underset{\sim}{M}$ is the centralizer of $\underset{\sim}{A}$ in $\underset{\sim}{G}$. We call $\underset{\sim}{A}$ a split component of $\underset{\sim}{P}$. Let N be the group of Ω-rational points of $\underset{\sim}{N}$. For any split component $\underset{\sim}{A}'$ of $\underset{\sim}{P}$ there exists a unique element $n \in N$ such that $\underset{\sim}{A}' = \underset{\sim}{A}^n$. Hence $\dim \underset{\sim}{A}$ depends only on $\underset{\sim}{P}$. We call it the parabolic rank of $\underset{\sim}{P}$ and denote it by $\operatorname{prk} \underset{\sim}{P}$.

Now let Ω be a $\mathfrak{p}$-adic field (i.e. a locally compact field with a non-trivial discrete valuation[1]). Let G denote the group of all Ω-rational points of $\underset{\sim}{G}$. By a parabolic subgroup (or cuspidal subgroup) P of G, we mean a subgroup of the form $P = G \cap \underset{\sim}{P}$, where $\underset{\sim}{P}$ is an Ω-parabolic subgroup of $\underset{\sim}{G}$. P determines $\underset{\sim}{P}$ completely. By a split component A of P, we mean a subgroup of the form $\underset{\sim}{A} \cap G$, where $\underset{\sim}{A}$ is a split component of $\underset{\sim}{P}$. $\underset{\sim}{A}$ is completely determined by A, since Ω is an infinite field. We write $\operatorname{prk} P = \dim \underset{\sim}{A}$. We call (P, A) a parabolic (or cuspidal) pair in G. Once A is fixed, we have the corresponding Levi Ω-decompositions $\underset{\sim}{P} = \underset{\sim}{M}\underset{\sim}{N}$ and $P = MN$ where $M = \underset{\sim}{M} \cap G$. We shall call N the unipotent radical of P.

---

[1] We always assume that the absolute value is given by $d(ax) = |a|dx$ $(a \in \Omega^*)$, $|0| = 0$, where dx denotes a Haar measure on the additive group of Ω.

§3. Supercuspidal representations.

Let G be as defined in §2. Then $G \subset GL(n, \Omega)$ for a suitable $n \geq 0$. $GL(n, \Omega)$, being an open subset of a vector space over Ω of dimension n^2, is a locally compact group. Since G is closed in $GL(n, \Omega)$, it is also locally compact. Moreover it is a separable and unimodular group. Let Z be the split component of G. Then Z satisfies the conditions of §1. Let f be a continuous function on G with compact support mod Z. For any parabolic subgroup P of G with unipotent radical N, put

$$f^P(x) = \int_N f(xn)dn \qquad (x \in G)$$

where dn is a fixed Haar measure on the unimodular group N. The integral exists since $Z \subset M$, M being a Levi subgroup of P such that $P = MN$.

Definition. Let f be a continuous function on G. f is said to be a supercusp form if

(i) Supp f is compact mod Z,

(ii) $f^P = 0$ for all parabolic subgroups $P \neq G$.

For $\chi \in \hat{Z}$, denote by ${}^\circ C_c(G, \chi)$ the space of supercusp forms in $C_c(G, \chi)$. Let π be an irreducible unitary representation of G on $\mathcal{H}$.

Definition. We call π supercuspidal if there exist $\phi, \psi \in \mathcal{H} - \{0\}$ such that the function $x \longmapsto (\phi, \pi(x)\psi)$ $(x \in G)$ is a supercusp form.

The concept of supercuspidal representation is clearly invariant under equivalence of representations. Let ${}^\circ\mathcal{E}(G)$ denote the set of all classes $\omega \in \mathcal{E}(G)$, which contain supercuspidal representations. Obviously ${}^\circ\mathcal{E}(G) \subset \mathcal{E}_2(G)$. For $\chi \in \hat{Z}$, put ${}^\circ\mathcal{E}(G, \chi) = {}^\circ\mathcal{E}(G) \cap \mathcal{E}(G, \chi)$.

Let G' be an open set in G and denote by $C_c^\infty(G')$ the space of all locally constant complex-valued functions on G' with compact support (no topology!). By a distribution on G' we mean simply a linear mapping of

$C_c^{\infty}(G')$ into the complex numbers.

Theorem 3. Let $\omega \in \mathcal{E}_2(G)$ and fix $\pi \in \omega$. For any $f \in C_c^{\infty}(G)$, the operator

$$\pi(f) = \int_G f(x)\pi(x)dx$$

is of the trace class (even of finite rank).

Since every neighborhood of 1 in G contains an open compact subgroup, this is an immediate consequence of the corollary of Theorem 2.

Define $\Theta_\pi(f) = \operatorname{tr} \pi(f)$ $(f \in C_c^{\infty}(G))$. Then Θ_π, called the character of π, is a distribution on G which depends only on the class ω of π. Hence we may denote it by Θ_ω. Then Θ_ω is invariant and the mapping $\omega \mapsto \Theta_\omega$ $(\omega \in \mathcal{E}_2(G))$ is injective.

§4. A conjecture.

Suppose, for the time being, that G is a real connected semisimple Lie group with finite center. Fix a maximal compact subgroup K of G. For $\omega \in \mathcal{E}(G)$ and $\underline{d} \in \mathcal{E}(K)$, $[\omega : \underline{d}]$ and $d(\underline{d})$ are as defined in §1. The following theorem is well-known.

Theorem. There exists an integer $N \geq 1$ such that

$$[\omega : \underline{d}] \leq N d(\underline{d})$$

for all $\omega \in \mathcal{E}(G)$ and $\underline{d} \in \mathcal{E}(K)$.

For linear G, one may choose $N = 1$ (cf. R. Godement, A theory of spherical functions I, Trans. A.M.S. 73 (1952), 496-556 for an instructive proof).

Now assume G $\mathfrak{p}$-adic (as defined in §2). This theorem will certainly hold for $\omega \in {}^{0}\mathcal{E}(G)$ (and all $\underline{d} \in \mathcal{E}(K)$, K being a maximal compact subgroup) if, in view of Theorem 2, the following conjecture is true.

Conjecture. $\inf_{\omega \in {}^{\circ}\mathcal{E}(G)} d(\omega) > 0.$

In view of results of Shalika for $SL(2, \Omega)$ (cf. P. J. Sally - J. A. Shalika, Proc. Nat. Ac. Sci. U.S.A. 61 (1968), pp. 1231-1237) one is inclined to believe that, at least for semisimple G, the formal degrees $d(\omega)$ $(\omega \in {}^{\circ}\mathcal{E}(G))$ are integers, up to a constant depending on the choice of the Haar measures. As shown by Shalika in case the residual characteristic of Ω is not 2, every $\pi \in {}^{\circ}\mathcal{E}(G)$ is induced by an irreducible representation τ of some maximal compact subgroup of $G (G = SL(2, \Omega))$. Assuming this, it is an easy exercise to prove that $d(\omega) = d(\underline{d})$, where $\underline{d}$ is the class of τ, provided the Haar measure on G is so normalized that the total measure of K is one.

The general case is, however, rather misty and no general result in this direction has been obtained. We shall return to this conjecture later on.

Part II. Existence of characters in the general case.

We keep the notations of Part I. We shall discuss the following theorem for reductive $\mathfrak{p}$-adic groups G.

Theorem 4. Let K be any open compact subgroup of G. Then

$$\sup_{\omega \in \mathcal{E}(G)} [\omega : \underline{d}] < +\infty \text{ for all } \underline{d} \in \mathcal{E}(K) .$$

A proof is not yet available for general reductive $\mathfrak{p}$-adic groups. We shall present a proof based on a conjecture about the dimension of spaces of a certain type of supercusp forms (to be defined in §3), which is suggested by the case of real groups.

The theorem has a few important consequences. First of all it implies that G is a type I-group, i.e. every factor representation of G is of type I (cf. R. Godement, Théorie des caractères, Ann. Math. 59 (1954), 47-85, Theorem 8). Furthermore it yields the existence of the character (as a distribution on G) for all $\omega \in \mathcal{E}(G)$. Up to now this is known in general only for the discrete series of G (cf. Part I).

§1. The Godement principle.

What follows is taken from a small, but important part of Godement's paper on spherical functions (cf. reference in Part I, §4).

Let A be an associative algebra over the complex numbers. For $a_1, \ldots, a_r \in A$ $(r \geq 1)$, define

$$[a_1, \ldots, a_r] = \sum_{\sigma} \varepsilon(\sigma) a_{\sigma(1)} a_{\sigma(2)} \cdots a_{\sigma(r)} ,$$

where the sum is taken over all permutations σ of the numbers 1, 2, ..., r and $\varepsilon(\sigma)$ denotes the sign of σ.

A is called r-abelian if $[a_1, \ldots, a_r] = 0$ for all r-tuples $a_1, \ldots, a_r \in A$. If A is r-abelian and $s \geq r$, then A is also s-abelian.

If dim $A = r$, then A is $(r+1)$-abelian.

Lemma 2. Let $r, s \geq 1$. Suppose for any a_{ij} $(1 \leq i \leq r,\ 1 \leq j \leq s)$,

$$[b_1, \ldots, b_r] = 0$$

where $b_i = [a_{i1}, a_{i2}, \ldots, a_{is}]$ $(1 \leq i \leq r)$. Then A is rs-abelian.

Let V be a complex vector space of dimension n.

Lemma 3. Let $A = \mathrm{End}(V)$. If A is r-abelian, then $r > n$.

To prove this lemma, observe that $[E_{1,2}, E_{2,3}, \ldots, E_{n-1,n}] = E_{1,n}$ where $E_{i,j}$ is the matrix $(\varepsilon_{k\ell})$ (w.r.t. some base of V) with $\varepsilon_{ij} = 1$, $\varepsilon_{k\ell} = 0$ otherwise.

Let again $A = \mathrm{End}(V)$, $\dim V = n$. Denote by $r(n)$ the smallest integer $r > 0$ such that $[X_1, \ldots, X_r] = 0$ for all $X_1, \ldots, X_r \in A$. One has $r(n+1) > r(n)$ (cf. e.g. J. Dixmier, Les C^*-algèbres et leurs représentations, Gauthier-Villars, Paris 1964, 3.6.2).

In this paragraph G is a locally compact, separable and unimodular group and K is a (non necessarily open) compact subgroup of G. We introduce two algebras of functions.

Let the Haar measure on K be normalized such that the total measure of K is one. Denote by τ a finite-dimensional continuous unitary representation of K on V and by $\check{\tau}$ the representation of K contragredient to τ, i.e. $\check{\tau}(k) = {}^t\tau(k^{-1})$ $(k \in K)$.

Let $C_c(G, \tau)$ be the convolution algebra of the continuous mappings $f : G \longrightarrow \mathrm{End}(V)$ with compact support and satisfying

$$f(kxk') = \tau(k)f(x)\tau(k') \qquad (k, k' \in K, x \in G) .$$

Now assume τ irreducible. Let $\underline{d}$ be the class of $\check{\tau}$. Let ξ be a unit vector in V. Define

$$e_{\underline{d}}(k) = d(\underline{d})\ \overline{(\xi,\ \tau(k)\xi)} \qquad (k \in K)\ .$$

Observe that $e_{\underline{d}} * e_{\underline{d}} = e_{\underline{d}}$.

Denote by $C_c(G, e_{\underline{d}})$ the convolution algebra of the continuous complex-valued functions f on G with compact support, which satisfy

$$e_{\underline{d}} * f * e_{\underline{d}} = f\ .$$

(The convolution product may be viewed as convolution of measures on G.) We have the following lemma.

Lemma 4. <u>Fix</u> $\underline{d} \in \mathcal{E}(K)$ <u>and</u> $\check{\tau} \in \underline{d}$. <u>The algebras</u> $C_c(G, \tau)$ <u>and</u> $C_c(G, e_{\underline{d}})$ <u>are isomorphic</u> (cf. G. van Dijk, Spherical functions on the $\mathfrak{p}$-adic group $\bar{P}GL(2)$, Proc. Amsterdam, Series A, <u>72</u>, 213-241, Theorem (1.1)).

The next lemma is only needed in a weaker form. As stated, it implies for real groups the well-known result of Part I, §4.

Lemma 5. <u>Fix</u> $\underline{d} \in \mathcal{E}(K)$ <u>and</u> $\check{\tau} \in \underline{d}$. <u>Then the following two statements are equivalent</u>.

(i) $\sup\limits_{\omega \in \mathcal{E}(G)} [\omega : \underline{d}] \leq n$,

(ii) $C_c(G, \tau)$ <u>is</u> $r(n)$-<u>abelian</u>.

<u>Proof</u>. Let V be the space of τ. Let ξ be a unit vector in V and put $e(k) = d(\underline{d})\ \overline{(\xi,\ \tau(k)\xi)}$ $(k \in K)$. From (i) it follows that $C_c(G, e)$ and hence $C_c(G, \tau)$ admits a complete set of representations of degree $\leq n$. So (ii) is obvious.

Now assume (ii) and switch again to $C_c(G, e)$. Fix $\omega \in \mathcal{E}(G)$ and $\pi \in \omega$. Denote by $\mathcal{H}$ the space of π and let $\mathcal{H}_e = \pi(e)(\mathcal{H})$. Let W be any finite-dimensional subspace of $\mathcal{H}_e$. Extend the elements of End(W) in the obvious way to $\mathcal{H}$. By von Neumann's density theorem every element of End(W) can be approximated by operators of the form $\pi(f)$ $(f \in C_c(G, e))$

in the strong operator topology of $\mathrm{End}(\mathcal{H})$. So $\mathrm{End}(W)$ is $r(n)$-abelian. By Lemma 3 we have $\dim W \leq r(n)$. Hence $\dim \mathcal{H}_e \leq r(n)$ and moreover $\mathrm{End}(\mathcal{H}_e)$ is $r(n)$-abelian. Therefore $\dim \mathcal{H}_e \leq r(n)$, since $r(n+1) > r(n)$. But this implies $[\omega : \underline{d}] \leq n$ and hence (i).

We call the assertion of Lemma 5 the Godement principle.

§2. A theorem of Bruhat and Tits.

Let G be a reductive $\mathfrak{p}$-adic group, as defined in Part I, §2. Observe that the following statements are equivalent.

a) For one, fixed open compact subgroup K_o of G, one has

$$\sup_{\omega \in \mathcal{E}(G)} [\omega : \underline{d}] < +\infty \text{ for all } \underline{d} \in \mathcal{E}(K_o) .$$

b) Let K be any open compact subgroup of G. Then $\sup_{\omega \in \mathcal{E}(G)} [\omega : \underline{d}] < +\infty$ for all $\underline{d} \in \mathcal{E}(K)$.

c) Let K be any open compact subgroup of G. Then $\sup_{\omega \in \mathcal{E}(G)} [\omega : \underline{d}_o] < +\infty$ where $\underline{d}_o$ = class of the identity representation of K.

In the discussion of Theorem 4 a particular K, given by Bruhat and Tits, will occur. Let us recall a recent theorem of them, which is basic not only here but sometimes even more in the next Parts.

First we introduce some more terminology. Let (P_i, A_i) $(i = 1, 2)$ be two cuspidal pairs in G. We write $(P_1, A_1) \succ (P_2, A_2)$ if $P_1 \supset P_2$ and $A_1 \subset A_2$. A cuspidal pair is called mincuspidal if it is minimal with respect to this partial order. Let $\mathcal{W}(A_1, A_2)$ denote the set of all bijections $s : A_1 \longrightarrow A_2$ with the following property. There is an element $y \in G$ such that $a^s = a^y$ $(= yay^{-1})$ for all $a \in A_1$. It is known that $\mathcal{W}(A_1, A_2)$ is a finite set. Fix $s \in \mathcal{W}(A_1, A_2)$. We say that $y \in G$ is a representative of s in G if $a^s = a^y$ for all $a \in A_1$. In case $A_1 = A_2 = A$ we write $\mathcal{W}(A)$ instead of $\mathcal{W}(A, A)$. Then $\mathcal{W}(A)$ is a finite group. Let (P_o, A_o) be a mincuspidal pair in G and $P_o = M_o N_o$ the corresponding Levi-decomposition. For any root α of (P_o, A_o), let ξ_α be the corresponding character of $\underset{\sim}{A}_o$. Let A_o^+ be the set of all points $a \in A_o$ where $|\xi_\alpha(a)| \geq 1$ for every root α of (P_o, A_o).

Theorem 5 (Bruhat and Tits). We can choose an open and compact subgroup K of G with the following properties.

(i) $G = K P_o$.

(ii) $G = K A_o^+ \omega_{M_o} K$, where ω_{M_o} is a finite subset of M_o.

(iii) Every element of $\mathcal{W}(A_o)$ has a representative in K.

(iv) If $(P, A) \succ (P_o, A_o)$ is a cuspidal pair and $P = MN$ the corresponding Levi-decomposition, then $P \cap K = (M \cap K)(N \cap K)$.

(v) Put $K_M = K \cap M$ and ${}^*P_o = M \cap P_o$. If we replace (G, P_o, A_o, K) by $(M, {}^*P_o, A_o, K_M)$, the above four conditions are again fulfilled.

Let P be any parabolic subgroup of G. There is a minimal parabolic subgroup contained in P. Since the minimal parabolic subgroups are Ω-conjugate to each other, we can find an element k in G, and by Theorem 5 (i), even in K, such that $P^k \supset P_o$. We then obviously have $G = K P$. Moreover we can choose a split component A of P in such a way that $(P^k, A^k) \succ (P_o, A_o)$. Let $P = MN$ be the corresponding Levi decomposition. Then $P^k = M^k N^k$ and by Theorem 5 (iv), $P^k \cap K = (M^k \cap K)(N^k \cap K)$, hence $P \cap K = (M \cap K)(N \cap K)$.

§3. Proof of Theorem 4 (based on Conjecture I).

Let G be a reductive $\mathfrak{p}$-adic group. We recall the conjecture stated in Part I, §4 and put:

Conjecture I. Let K be any open compact subgroup of G and let ${}^oC_c(G/\!/K, \chi)$ $(\chi \in \hat{Z})$ denote the space of all functions in ${}^oC_c(G, \chi)$ which are constant on double cosets w.r.t. K. Then

$$\sup_{\chi \in \hat{Z}} \dim {}^oC_c(G/\!/K, \chi) < +\infty .$$

Conjecture II. $\inf_{\omega \in {}^o\mathcal{E}(G)} d(\omega) > 0.$

In Part III we shall prove that Conjecture II implies Conjecture I.

Proposition 1. Assuming that Conjecture I is true, Theorem 4 applies to all reductive $\mathfrak{p}$-adic groups G.

Before giving the proof, we make first the following observations. Let (P, A) be a cuspidal pair and $P = MN$ the corresponding Levi-decomposition. Let K be the open compact subgroup of Theorem 5. We have the following integration formulae.

Let $d_r p$ denote a right Haar measure on P and define the function δ_P on P by the relation $d_r(qp) = \delta_P(q) d_r p$ $(q \in P)$. Then, provided the Haar measures are suitably normalized,

$$\int_G f(x)dx = \int_P \int_K f(kp)dk d_r p \qquad (f \in C_c(G)) .$$

This formula is well-known for subgroups K and P such that $G = KP$, $K \cap P = \{e\}$, but it is clear that it persists to be true if $K \cap P$ is compact. Furthermore $d_r p = dndm$, corresponding to the decomposition $P = NM$ and $d_r p = \delta_P(m)dmdn$ corresponding to $P = MN$.

Proof of Proposition 1. We proceed by induction on the semi-simple Ω-rank of G (= dimension of a maximal Ω-split torus in the derived group of G).

a) ssrk $G = 0$. Then G/Z is compact. Let K be any open compact subgroup of G. The group KZ is open in G and $[G : KZ] < +\infty$. An easy observation gives $\sup_{\omega \in \mathcal{E}(G)} [\omega : \underline{d}] \leq [G : KZ]$ for all $\underline{d} \in \mathcal{E}(K)$.

b) ssrk $G > 0$. Choose K and (P_o, A_o) as in Theorem 5. Fix a cuspidal pair (P, A) with $P \neq G$, such that $(P, A) \succ (P_o, A_o)$. Let $P = MN$ be the corresponding Levi-decomposition. Then ssrk M < ssrk G. Fix $\underline{d} \in \mathcal{E}(K)$ and $\tau \in \underline{d}$. Let V be the space of τ. Denote by V_P the subspace of all $v \in V$ such that $\tau(n)v = v$ for all $n \in N \cap K$ and let E_P be the orthogonal projection of V on V_P. Put

$$\tau_M(m) = \tau(m)E_P \qquad (m \in K \cap M) .$$

Then τ_M may be regarded as a representation of $K \cap M$ on V_P. Consider the algebras $C_c(G, \tau)$ and $C_c(M, \tau_M)$. For any $f \in C_c(G, \tau)$ put

$$f^{(P)}(m) = \delta_P(m)^{1/2} \int_N f(mn)E_P dn \qquad (m \in M) ,$$

where dn is a Haar measure on N. Then $f^{(P)} \in C_c(M, \tau_M)$ and the mapping $\mu_P : f \longmapsto f^{(P)}$ is a homomorphism of $C_c(G, \tau)$ into $C_c(M, \tau_M)$. Now $C_c(M, \tau_M) \subset C_c(M, \tau_o) \otimes \mathrm{End}(V_P)$, where τ_o is the identity representation of certain compact open subgroup of M, contained in $K \cap M$. Applying the induction hypothesis and Lemma 5, we conclude that $C_c(M, \tau_M)$ is r-abelian for some $r = r(P, A)$. Choose $s \geq r(P, A)$. Let $f_i \in C_c(G, \tau)$ $(1 \leq i \leq s)$ and put $\phi = [f_1, \ldots, f_s]$. Then $\mu_P(\phi) = 0$. For s sufficiently large we even have $\mu_P(\phi) = 0$ for all cuspidal pairs (P, A) with $P \neq G$, dominating (P_o, A_o). Now let P' be any parabolic subgroup, $P' \neq G$. There exists $k \in K$ such that $P'^k \supset P_o$. Since $\phi(kxk^{-1}) = \tau(k)\phi(x)\tau(k^{-1})$ $(x \in G)$, it is clear that $\mu_{P'}(\phi) = 0$. This implies that ϕ is a supercusp form. We have proved: there exists $s \geq 1$ such that for all $f_1, \ldots, f_s \in C_c(G, \tau)$, $\phi = [f_1, \ldots, f_s]$ belongs to ${}^oC_c(G, \tau)$. Now Conjecture I implies that ${}^oC_c(G, \tau)$ is p-abelian for some large p (we need only this implication!).

Indeed, let π be a χ-representation $(\chi \in \hat{Z})$. Let $\phi \in C_c(G)$. We have as before $\pi(\phi) = \pi_*(\phi_{\bar{\chi}})$ where $\phi_{\bar{\chi}}(x) = \int_Z \phi(xz)\chi(z)dz$ $(x \in G)$. Observe that

${}^oC_c(G, \tau) \subset {}^oC_c(G/\!/K_o) \otimes \mathrm{End}(V)$ for some open compact subgroup $K_o \subset G$. If $\phi_1, \ldots, \phi_p \in {}^oC_c(G/\!/K_o)$, then $\pi([\phi_1, \ldots, \phi_p]) = [\pi_*(\phi_{1,\bar{\chi}}), \ldots, \pi_*(\phi_{p,\bar{\chi}})] = 0$ as soon as

$$p > \sup_{\chi \in \hat{Z}} \dim {}^oC_c(G/\!/K_o, \chi)$$

for all χ-representations π and all $\chi \in \hat{Z}$. Therefore ${}^oC_c(G, \tau)$ is p-abelian.

Now take $f_{ij} \in C_c(G, \tau)$ $(1 \leq i \leq p,\ 1 \leq j \leq s)$. Put

$\phi_i = [f_{i_1}, \ldots, f_{i_s}]$ $(1 \leq i \leq p)$. Then $\phi_i \in {}^{o}C_c(G, \tau)$ for all i and $[\phi_1, \ldots, \phi_p] = 0$. Hence, by Lemma 2, $C_c(G, \tau)$ is ps-abelian. The proposition now follows immediately from Lemma 5. This completes the proof.

Part III. Supercusp forms and supercuspidal representations.

§1. The space generated by a supercusp form.

Let G be a reductive $\mathfrak{p}$-adic group. Choose a mincuspidal pair (P_o, A_o) and let K be the corresponding open compact subgroup given by Theorem 5 with the properties listed there. In particular

$$G = K\omega_{M_o} A_o^{+} K ,$$

where ω_{M_o} is a finite subset of M_o ($P_o = M_o N_o$).

Let $\Sigma = \Sigma(P_o/A_o)$ be the set of roots of (P_o, A_o). For $\alpha \in \Sigma$ let $\mathfrak{n}_\alpha$ denote the root space of α in the Lie algebra of G. Let $\Sigma^o = \Sigma^o(P_o/A_o)$ be the set of simple roots of (P_o, A_o). We denote by ξ_α the character of $\underset{\sim}{A}_o$ corresponding to the root α.

For any cuspidal pair $(P, A) \succ (P_o, A_o)$, $P = MN$, there exists a subset $F \subset \Sigma^o$ such that $\underset{\sim}{A} = (\bigcap_{\alpha \in F} \ker \xi_\alpha)^o$ and the Lie algebra of N is $\Sigma \mathfrak{n}_\alpha$, the sum going over all (positive) roots that are not linear combinations of elements in F. We write $(P, A) = (P_o, A_o)_F$.

From now on, let F denote a finite subset of $\mathcal{E}(K)$. Let π be a continuous unitary representation of G. Define

$$\alpha_F = \sum_{\underline{d} \in F} \alpha_{\underline{d}}$$

where $\alpha_{\underline{d}}$ is as defined in Part I, §1. We extend α_F to G by putting $\alpha_F(x) = 0$ if $x \notin K$. Then $\alpha_F \in C_c^\infty(G)$.

Denote by $E = E(F, \pi)$ the projection $\pi(\alpha_F)$. Let $\mathcal{H}_\pi$ be the space of π. We write $\mathcal{H}_\pi(F)$ for $E(F, \pi)\mathcal{H}_\pi$. Clearly

$$\mathcal{H}_\pi(F) = \bigoplus_{\underline{d} \in F} \mathcal{H}_\pi(\underline{d})$$

where $\mathcal{H}_\pi(\underline{d}) = \pi(\alpha_{\underline{d}})\mathcal{H}_\pi$.

By $C_c(G, \alpha_F)$ we denote the convolution algebra of the complex-valued continuous functions f with compact support, satisfying $\alpha_F * f * \alpha_F = f$.

Lemma 6. *Fix* $\alpha \in \Sigma^o$, $\chi \in \hat{Z}$, $f \in {}^oC_c(G, \overline{\chi})$ *and a finite subset* $F \subset \mathcal{E}(K)$. *Then there exists* $t \geq 1$ *with the following property. Let* π *be a* χ-*representation and put* $E = E(F, \pi)$. *Then*

$$\pi_*(f)\pi(m_o a_o)E = 0$$

for $m_o \in \omega_{M_o}$, $a_o \in A_o^+$ *unless* $|\xi_\alpha(a_o)| \leq t$.

Proof. Let F_α denote the complement of $\{\alpha\}$ in Σ^o. Put $(P, A) = (P_o, A_o)_{F_\alpha}$. Let $P = MN = NM$ be the corresponding Levi-decomposition. Notice that $Z \subset M$. We have

$$\pi_*(f) = \int_{G/Z} f(x)\pi(x)dx^* = \int_K \int_N \int_{M/Z} f(knm)\pi(knm)dkdndm^*$$

and

$$\pi_*(f)\pi(m_o a_o) = \int_K \int_N \int_{M/Z} f(knm)\pi(kmm_o a_o)\pi(n^{(mm_o a_o)^{-1}})dkdndm^* .$$

If $knm \in \operatorname{Supp} f$, n and m^* remain bounded, so $n_1 = n^{(mm_o)^{-1}} \in N$ remains bounded ($m_o \in \omega_{M_o}$), hence

$$n^{(mm_o a_o)^{-1}} = n_1^{a_o^{-1}} \in K_F = \bigcap_{\underline{d} \in F} \ker \underline{d} ,$$

a (small) open compact subgroup of G, provided $|\xi_\alpha(a_o)| \geq t$ for some (large) $t \geq 1$. But then $\pi(n_1^{a_o^{-1}})E = E$. Therefore, if t is sufficiently large,

$$\pi_*(f)\pi(m_o a_o)E = \int_K \int_N \int_{M/Z} f(knm)\pi(kmm_o a_o)Edkdndm^* .$$

But

$$\int_N f(knm)dn = \int_N f(kmn^{m^{-1}})dn = \delta_P(m)\int_N f(kmn)dn = 0$$

since f is a supercusp form, provided $F_\alpha \neq \phi$.

If $F_\alpha = \phi$, the lemma is clearly true. We conclude

$$\pi_*(f)\pi(m_o a_o)E = 0$$

for $m_o \in \omega_{M_o}$ and $a_o \in A_o^+$ such that $|\xi_\alpha(a_o)| \geq t$. This proves the lemma.

Corollary. There exists $t \geq 1$ such that

$$\pi_*(f)\pi(m_o a_o)E = 0$$

for $m_o \in \omega_{M_o}$, $a_o \in A_o^+$ unless $\sup_{\alpha \in \Sigma_o} |\xi_\alpha(a_o)| \leq t$.

Put $A_o^+(t)$ = set of all $a \in A_o^+$ such that $\sup_{\alpha \in \Sigma_o} |\xi_\alpha(a)| \leq t$. This set is compact mod Z.

Lemma 7. Fix $\chi \in \hat{Z}$, $f \in {}^oC_c^\infty(G, \overline{\chi})$ and $F \subset \mathcal{E}(K)$. There exists a subset ω of G, which is compact mod Z, with the following property. If π is a χ-representation and $E = E(F, \pi)$, then

$$\pi_*(f)\pi(x)E = E\pi(x)\pi_*(f) = 0$$

unless $x \in \omega$ $(x \in G)$.

Proof. The function f is K-finite. Let $f_1, \ldots, f_r$ be a base for the space of functions spanned by the right translates of f under K. Choose $t \geq 1$ in the preceding corollary so large that

$$\pi_*(f)\pi(m_o a_o)E = 0$$

for $1 \leq i \leq r$, $m_o \in \omega_{M_o}$, $a_o \in A_o^+$ unless $a_o \in A_o^+(t)$.

Define $\omega_1 = K\omega_{M_o} A^+(t)K$. ω_1 is compact mod Z. Let $x \notin \omega_1$. Since

$G = K\omega_{M_o} A_o^+ K$, we have

$$x = k_1 m_o a_o k_2 \qquad (k_1, k_2 \in K, m_o \in \omega_{M_o}, a_o \in A_o^+)$$

with $a_o \in {}^c A^+(t)$. Hence

$$\pi_*(f)\pi(x)E = \pi_*(f)\pi(k_1)\pi(m_o a_o)E\pi(k_2)$$
$$= \pi_*(g)\pi(m_o a_o)E\pi(k_2) = 0$$

where $g(x) = f(xk_1^{-1})$ $(x \in G)$.

Furthermore $E\pi(x)\pi_*(f) = 0$ for $x \in {}^c\omega_2^{-1}$, where ω_2 is chosen in the same way as ω_1 for $\tilde{f}$ instead of f. Here $\tilde{f}(x) = f(x^{-1})$ $(x \in G)$. This is easily seen by taking the adjoint of $E\pi(x)\pi_*(f)$. Put $\omega = \omega_1 \cup \omega_2^{-1}$. Then $\pi_*(f)\pi(x)E = E\pi(x)\pi_*(f) = 0$ for $x \in {}^c\omega$. This completes the proof.

Denote by ${}^o C_c(G, \chi, \alpha_F)$ the space of functions $f \in {}^o C_c(G, \chi)$ which satisfy $\alpha_F * f * \alpha_F = f$, F being a fixed finite subset of $\mathcal{E}(K)$.

Lemma 8. <u>The elements of</u> ${}^o C_c(G, \bar{\chi}, \alpha_F)$ <u>are Hecke-finite: Fix</u> $f \in {}^o C_c(G, \bar{\chi}, \alpha_F)$. <u>Let</u> J_f <u>be the space spanned by all functions of the form</u> $\alpha * f * \beta$ $(\alpha, \beta \in C_c(G, \alpha_F))$. <u>Then</u> $\dim J_f < +\infty$.

<u>Proof.</u> First of all $\alpha * f = \alpha_{\bar{\chi}} *' f$ for all $\alpha \in C_c(G, \alpha_F)$, where $*'$ denotes convolution on G/Z and $\alpha_{\bar{\chi}}$ is defined by

$$\alpha_{\bar{\chi}}(x) = \int_Z \alpha(xz)\bar{\chi}(z^{-1})dz \qquad (x \in G).$$

Similarly $f * \beta = f *' \beta_{\bar{\chi}}$.

Let π be a χ-representation. Then we have for $\alpha, \beta \in C_c(G, \alpha_F)$,

$$\pi_*(\alpha * f * \beta) = \pi(\alpha)\pi_*(f)\pi(\beta) = E\pi(\alpha)\pi_*(f)\pi(\beta)E = \int_G \int_G \alpha(x)E\pi(x)\pi_*(f)\pi(y)E\beta(y)dxdy.$$

Choose a compact set ω such that

$$\pi_*(f)\pi(x)E = E\pi(x)\pi_*(f) = 0$$

for x outside ωZ (Lemma 7). Let ϕ be the characteristic function of the set $K\omega ZK$. Then obviously

$$\pi_*(\alpha * f * \beta) = \pi_*((\phi\alpha) * f * (\phi\beta))$$

for all $\alpha, \beta \in C_c(G, \alpha_F)$.

Now take for π the left regular representation of G on $L_2(G, \chi)$. Then it follows that $\alpha * f * \beta = (\phi\alpha) * f * (\phi\beta) = (\phi\alpha)_{\bar{\chi}} *' f *' (\phi\beta)_{\bar{\chi}}$ for all $\alpha, \beta \in C_c(G, \alpha_F)$. Notice that $\mathrm{Supp}(\phi\alpha)_{\bar{\chi}} \subset K\omega ZK$ and that $(\phi\alpha)_{\bar{\chi}}$ is constant on double cosets with respect to K_F, where $K_F = \bigcap_{\underline{d} \in F} \ker \underline{d}$ for all $\alpha \in C_c(G, \alpha_F)$. Therefore the functions $(\phi\alpha)_{\bar{\chi}}$ and $(\phi\beta)_{\bar{\chi}}$ remain in a finite-dimensional space: $\dim J_f < +\infty$. The lemma follows.

The following theorem has its real analogue. The proofs are mutatis mutandis the same (cf. Harish-Chandra, Discrete series for semisimple Lie groups II, Acta Math. 116 (1966), 1-111, Lemma 77).

Theorem 6. Fix $f \in {}^{\circ}C_c^{\infty}(G, \bar{\chi})$, $f \neq 0$. Let λ denote the left regular representation of G on $L_2(G, \bar{\chi})$. Let $\mathcal{H}_f$ be the smallest closed subspace of $\mathcal{H} = L_2(G, \bar{\chi})$, which is stable under λ and which contains f. Then $\mathcal{H}_f = \sum_{1 \leq i \leq r} V_i$, where V_i are closed, mutually orthogonal, λ-stable irreducible subspaces of $\mathcal{H}$. Let λ_i = restriction of λ on V_i. Then λ_i is a supercuspidal χ-representation.

Proof. There is an open compact subgroup K_o of G, contained in K such that f is constant on double cosets with respect to K_o. From this follows the existence of a finite subset $F \subset \mathcal{E}(K)$ such that

$$\alpha_F * f * \alpha_F = f \ .$$

(Consider the irreducible components of $\operatorname{ind}_{K_o \uparrow K} 1$.) Hence $f \in {}^{\circ}C_c(G, \bar{\chi}, \alpha_F)$.

Put $E = \lambda(\alpha_F)$. Consider $E(\mathcal{H}_f)$. The functions of the form $\alpha * f$ $(\alpha \in C_c(G))$ are dense in $\mathcal{H}_f$. Now

$$E(\alpha * f) = \alpha_F * \alpha * \alpha_F * f = \alpha' * f$$

where $\alpha' = \alpha_F * \alpha * \alpha_F$, so $\alpha' \in C_c(G, \alpha_F)$. Hence the functions of the form $\alpha * f * \alpha_F$ $(\alpha \in C_c(G, \alpha_F))$ are dense in $E(\mathcal{H}_f)$. By Lemma 8, $\dim E(\mathcal{H}_f) < +\infty$.

Let V be any non-zero closed, λ-stable subspace of $\mathcal{H}_f$. We claim that $E(V) \neq \{0\}$. For otherwise, suppose $E(V) = \{0\}$. Let U be the orthogonal complement of V in $\mathcal{H}_f$. Since $f = Ef$, f is orthogonal to V, hence $f \in U$. But U is also λ-stable and closed, hence $\mathcal{H}_f \subset U$. Therefore $V = \{0\}$, contradicting our hypothesis.

Let V_i $(1 \leq i \leq p)$ be a finite set of mutually orthogonal, closed, non-zero subspaces of $\mathcal{H}_f$, which are stable under λ. Since $E(V_i) \neq \{0\}$ for all i, we have $p \leq \dim E(\mathcal{H}_f) < +\infty$. Therefore if p is as large as possible, we get an irreducible decomposition.

Let λ_i denote the restriction of λ on V_i. Choose non-zero vectors ϕ, ψ in $E(V_i) \subset V_i$. Since $\phi, \psi \in E(\mathcal{H}_f)$ and $\dim E(\mathcal{H}_f) < +\infty$, we conclude from the above observations that $\phi, \psi \in {}^oC_c(G, \bar{\chi}, \alpha_F)$. We get $\theta(x) = (\phi, \lambda_i(x)\psi) = \int_{G/Z} \overline{\phi(y)}\psi(y^{-1}x)dy^*$ and, if P is a parabolic subgroup of G, $P \neq G$, N its unipotent radical, then, by interchanging integrals

$$\int_N \theta(xn)dn = \int_{G/Z} dy^* \int_N \overline{\phi(y)}\psi(n^{-1}x^{-1}y)dn$$

and

$$\int_N \psi(n^{-1}x^{-1}y)dn = \int_{N^{x^{-1}y}} \psi(y^{-1}xn)dn = 0 .$$

Hence λ_i is supercuspidal. Clearly λ_i is a χ-representation. This proves the theorem.

§2. Some consequences.

We come to the proof of the following proposition.

Proposition 2. Conjecture II implies Conjecture I.

Actually we shall prove the following stronger result. Let G be a reductive $\mathfrak{p}$-adic group.

Proposition 2'. If Conjecture II is true for G, then Conjecture I is.

Proof. Let K_o be any open compact subgroup of G and let $\chi \in \hat{Z}$. We can choose a finite subset $F \subset \mathcal{E}(K)$ such that

$$^{o}C_c(G/\!/K_o, \chi) \subset {}^{o}C_c(G, \chi, \alpha_F) .$$

We shall show $\sup_{\chi \in \hat{Z}} \dim {}^{o}C_c(G, \chi, \alpha_F) < +\infty$ for all finite subsets $F \subset \mathcal{E}(K)$.

Fix F and let $\omega \in \mathcal{E}(G)$. Put $[\omega : F] = \sum_{d \in F} [\omega : \underline{d}]$. Denote by $\mathcal{E}_F(G)$ the set of all $\omega \in \mathcal{E}(G)$ such that $[\omega : F] > 0$ and put ${}^{o}\mathcal{E}_F(G, \chi) = \mathcal{E}_F(G) \cap {}^{o}\mathcal{E}(G, \chi)$.

For $\omega \in \mathcal{E}_2(G, \chi)$, let $\mathcal{H}(\omega, F)$ be the space spanned by the matrix coefficients $x \longmapsto (\phi, \pi(x)\psi)$ where $\phi, \psi \in \mathcal{H}_\pi(F)$ and π is any representation in the class ω. By Theorem 2 we have $\dim \mathcal{H}(\omega, F) < +\infty$. Observe that $\mathcal{H}(\omega, F) \subset L_2(G, \chi)$. We claim ${}^{o}C_c(G, \chi, \alpha_F) = \bigoplus_{\omega \in {}^{o}\mathcal{E}_F(G, \chi)} \mathcal{H}(\omega, F)$. For K-finite vectors $\phi, \psi \in \mathcal{H}_\pi$ $(\pi \in \omega \in {}^{o}\mathcal{E}_F(G, \chi))$, $x \longmapsto (\phi, \pi(x)\psi)$ is a supercusp form. Hence $\mathcal{H}(\omega, F) \subset {}^{o}C_c(G, \chi, \alpha_F)$ for all $\omega \in {}^{o}\mathcal{E}_F(G, \chi)$. On the other hand we have for $f \in {}^{o}C_c(G, \chi, \alpha_F)$, with the notations of Theorem 6, $\mathcal{H}_f = \sum_{1 \le i \le p} V_i$, where V_i are closed, mutually orthogonal, λ-stable irreducible subspaces of $L_2(G, \chi)$. Let λ_i be the restriction of λ on V_i. Denote furthermore by E_i the orthogonal projection of $L_2(G, \chi)$ on V_i. Put $E = \lambda(\alpha_F)$. We have

$$f(y) = f * \alpha_F(y) = (\alpha_{F,\chi}, \lambda(y^{-1}f)) = \sum_{1 \le i \le r} (E_i \alpha_{F,\chi}, \lambda_i(y^{-1})E_i f) \qquad (y \in G) ,$$

where $a_{F,\chi}(x) = \int_Z a_F(xz)\chi(z^{-1})dz \quad (x \in G)$. Let ω_i be the class of the representation contragredient to λ_i. Since ω_i is supercuspidal and $EE_i f = E_i f$, $EE_i a_{F,\chi} = E_i a_{F,\chi}$ for $1 \leq i \leq r$, we get easily $f \in \sum_{1\leq i\leq r} \mathcal{H}(\omega_i, F)$. Therefore $\dim {}^oC_c(G, \chi, a_F) = \sum_{\omega \in {}^o\mathcal{E}_F(G,\chi)} \dim \mathcal{H}(\omega, F)$ and

$$\sum_{\omega \in {}^o\mathcal{E}_F(G,\chi)} \dim \mathcal{H}(\omega, F) = \sum_{\underline{d} \in F} \sum_{\omega \in {}^o\mathcal{E}_{\underline{d}}(G,\chi)} \dim \mathcal{H}(\omega, \underline{d}) = \sum_{\underline{d} \in F} \sum_{\omega \in {}^o\mathcal{E}_{\underline{d}}(G,\chi)} [\omega : \underline{d}]^2 d(\underline{d})^2 .$$

Now applying Conjecture II we get by Theorem 2,

$$d(\underline{d}) \geq \sum_{\omega \in \mathcal{E}_2(G,\chi)} d(\omega)[\omega : \underline{d}] \geq \sum_{\omega \in {}^o\mathcal{E}(G,\chi)} d(\omega)[\omega : \underline{d}] \geq c \cdot \sum_{\omega \in {}^o\mathcal{E}_{\underline{d}}(G,\chi)} [\omega : \underline{d}]$$

for all $\underline{d} \in \mathcal{E}(K)$, where c is a positive constant, not depending on F and χ. We conclude

$$\dim {}^oC_c(G, \chi, a_F) \leq \sum_{\underline{d} \in F} c^{-2} d(\underline{d})^4$$

for all F and χ. We have proved Conjecture I.

Due to the nice properties of K, listed in Theorem 5, we have the following lemma. Fix $\chi \in \hat{Z}$ and denote by ${}^oC(K, \chi)$ the space of all $f \in {}^oC_c(G, \chi)$ whose support is contained in the group KZ.

Lemma 9. <u>In order that</u> $f \in C_c(G, \chi)$ <u>with</u> $\operatorname{Supp} f \subset KZ$ <u>will be a supercusp form, it is necessary and sufficient that</u>

$$\int_{N\cap K} f(kn)dn = 0 \qquad (k \in K)$$

<u>for all parabolic subgroups</u> $P \neq G$ (N = <u>unipotent radical of</u> P).

<u>Proof</u>. Let P be any parabolic subgroup of G with unipotent radical N.

By Theorem 5 we can choose a Levi-component M of P in such a way that $P \cap K = (M \cap K)(N \cap K)$. Furthermore $P = MN$ and $G = KP$. Since $Z \subset M$, we have $N \cap K = N \cap KZ$ and the 'necessity' follows. Conversely, assume $\int_{N\cap K} f(kn)dn = 0$ $(k \in K)$ for all $P \neq G$. This remains true for $k \in KZ$, since f is a χ-function. Then

$$\int_N f(xn)dn = \int_N f(kmn'n)dn = \int_N f(kmn) = 0 \qquad (x = kmn') .$$

Indeed, assuming $kmn \in KZ$, we get $zmn \in K$ for some $z \in Z$, hence since $Z \subset M$, $zm \in K$. Moreover, if $m \in KZ$,

$$\int_N f(kmn)dn = \int_{N\cap K} f(kmn)dn = 0 .$$

Consequently f is a supercusp form. This proves the lemma.

It is clear from the lemma that one can characterize ${}^0C(K, \chi)$ completely in terms of K.

Assume for simplicity prk $G = 0$. Fix $\underline{d} \in \mathcal{E}(K)$ and let $\xi_{\underline{d}}$ be its character. Denote by ${}^0C(K)$ the space of functions $f \in {}^0C_c(G)$ with Supp $f \subset K$.

Definition. <u>We call</u> $\underline{d}$ <u>cuspidal if</u> $\xi_{\underline{d}}$ <u>is in</u> ${}^0C(K)$.

The matrix coefficients of cuspidal representations $\underline{d}$ in $\mathcal{E}(K)$ are all in ${}^0C(K)$, since they are linear combinations of left and right translates of $\xi_{\underline{d}}$. Fix $\tau \in \underline{d}$. Let V be the space of τ. Choose a unit vector $\xi \in V$ and put $e_{\underline{d}}(k) = d(\underline{d})\,\overline{(\xi, \tau(k)\xi)}$ $(k \in K)$. Consider the induced unitary representation

$$\operatorname{ind}_{K\uparrow G} \check{\tau}$$

It is known that the commuting (von Neumann-) algebra of this representation is generated by the right convolution operators with elements of $C_c(G, \tau)$. By Lemma 4 this algebra is isomorphic with the algebra $C_c(G, e_{\underline{d}})$, which is finite-dimensional by Lemma 8, since $e_{\underline{d}} * \alpha_{\underline{d}} = e_{\underline{d}}$. Hence $\operatorname{ind}_{K\uparrow G} \check{\tau}$ splits up into

finitely many irreducible unitary representations which are clearly super-cuspidal.

Mautner (Spherical functions over $\mathfrak{p}$-adic fields II, Amer. J. Math. 86, 171-200 (1964), §9) first observed this method for obtaining supercuspidal representations in case G = PGL(2). The results of Shalika were already mentioned in Part I, §4. It is clear that one is led by this method to find the cuspidal representations of finite reductive algebraic groups, unfortunately not only defined over a field, but also over a (local) ring. Except for the group SL(2, Ω), it is not known if the above method yields all supercuspidal representations of G.

Part IV. The space $\mathcal{A}(G, \tau)$.

Another conjecture will be introduced in this Part. If this should be true much of the theory of real groups could be imitated in the $\mathfrak{p}$-adic case, as we shall show.

§1. Conjecture III.

Let V be a finite-dimensional complex Hilbert space and let τ be a unitary representation of K (given by Bruhat-Tits) on V. Denote by $H(\tau)$ what was formerly called $C_c(G, \tau)$, the convolution algebra of the mappings $\beta : G \longrightarrow \mathrm{End}(V)$ with compact support, satisfying

$$\beta(k_1xk_2) = \tau(k_1)\beta(x)\tau(k_2) \qquad (k_1, k_2 \in K, x \in G) .$$

Let (P, A) be any parabolic pair in G, $P = MN$ the corresponding Levi-decomposition. Put

$$V_P = \text{subspace of all } v \in V \text{ such that } \tau(n)v = v \text{ for all } n \in N \cap K .$$

Let E_P be the orthogonal projection of V on V_P. Put

$$\tau_M(m) = \tau(m)E_P \qquad (m \in K \cap M) .$$

Then τ_M may be regarded as a representation of $K \cap M$ on V_P and so we can consider the algebra $H(\tau_M)$. For $\beta \in H(\tau)$ put

$$\beta^{(P)}(m) = \delta_P(m)^{1/2} \int_N \beta(mn)dn \qquad (m \in M) .$$

Then $\beta^{(P)}(m) = E_P\beta^{(P)}(m)E_P$ for all $m \in M$. Moreover $\beta^{(P)} \in H(\tau_M)$. The mapping $\mu_P : \beta \longmapsto \beta^{(P)}$ is actually a homomorphism of $H(\tau)$ into $H(\tau_M)$.

Conjecture III. $H(\tau_M)$ <u>is a finite right-module over</u> $\mu_P(H(\tau))$, i.e. <u>there exist</u> $p \geq 1$ <u>and</u> $\alpha_1, \ldots, \alpha_p \in H(\tau_M)$ <u>such that</u>

$$H(\tau_M) = \sum_{1 \leq i \leq p} \alpha_i * \mu_P(H(\tau)) .$$

Observe that it is enough to state the conjecture for the parabolic pairs (P, A) dominating a fixed mincuspidal pair (P_o, A_o). The real 'analogue' of the conjecture reads as follows.

Let $\mathfrak{z}, \mathfrak{z}_M$ be the centers of the universal enveloping algebras of the complexification of the Lie algebras of G, M respectively. There is an (injective) homomorphism $\mu : \mathfrak{z} \longrightarrow \mathfrak{z}_M$ such that $\mathfrak{z}_M$ can be viewed as a free $\mathfrak{z}$-module of finite rank (cf. Harish-Chandra, Invariant eigendistributions on a semisimple Lie group, Trans. A.M.S. 119 (1965), 457-508, Lemma 21).

About the proof of the conjecture, one special case is known to be true, Satake has proved it in case char $\Omega = 0$, τ = identity representation of K and P is minimal.[+] In this case $H(\tau_M)$ is a free module of finite rank over $\mu_P(H(\tau))$ (cf. I. Satake, Theory of spherical functions on reductive algebraic groups over a $\mathfrak{p}$-adic field, Publ. Math. I.H.E.S. 18, 5-70, Ch. II),

§2. The space $\mathcal{A}(G, \tau)$.

Let G and τ be as in §1. We denote by $\mathcal{A}(G, \tau)$ the space of all functions $f : G \longrightarrow \mathrm{End}(V)$ such that

(i) $f(k_1 x k_2) = \tau(k_1) f(x) \tau(k_2)$ $(k_1, k_2 \in K, x \in G)$

(ii) f is left $H(\tau)$-finite.

This space has its real analogue (cf. Harish-Chandra, Harmonic analysis on semisimple Lie groups, Colloquium Lectures, University of Oregon, A.M.S. 1969, §9). We shall prove the same kind of results for this space in the $\mathfrak{p}$-adic case under the assumption that Conjecture III is true.

Let (P, A) be a cuspidal pair, P = MN the corresponding Levi-decomposition, $\Sigma = \Sigma(P, A)$ the roots of (P, A) and Σ^o the set of simple roots of Σ. For $\alpha \in \Sigma$ let ξ_α be the corresponding character of A. Put $\gamma_P(a) = \inf_{\alpha \in \Sigma^o} |\xi_\alpha(a)|$, and for $c > 0$,

$$A^+(c) = \text{set of all } a \in A \text{ such that } \gamma_P(a) \geq c \ .$$

We write A^+ for $A^+(1)$.

+) In addition, $\underset{\sim}{G}$ has to be assumed simply connected.

Definition. **Let ϕ be a continuous function on M. We write $\phi \underset{P}{\approx} 0$ if, given a compact set ω in M, we can choose $c > 0$ such that $\phi = 0$ on $\omega A^+(c)$.**

We shall prove the following theorems:

Theorem 7 (based on Conjecture III). Fix $f \in \mathcal{A}(G, \tau)$ and put $f_{(P)}(m) = \delta_P(m)^{1/2} f(m)$ $(m \in M)$. There exists a unique element $f_P \in \mathcal{A}(M, \tau_M)$ such that

$$f_{(P)} - f_P \underset{P}{\approx} 0 .$$

Theorem 8 (based on Conjecture III). Fix $f \in \mathcal{A}(G, \tau)$ and assume $f_P = 0$ for all $P \neq G$. Then f is a supercusp form.

The proofs require a rather long preparation. In particular we need the $\mathfrak{p}$-adic analogue of (Harish-Chandra, Discrete series for semisimple Lie groups II, Acta Math. 116 (1966), 1-111, Part II).

Lemma 10. Let $\phi \in \mathcal{A}(M, \tau_M)$ and suppose $\phi \underset{P}{\approx} 0$. Then $\phi = 0$.

Proof. Denote by Φ the finite-dimensional space consisting of all functions of the form $\alpha * \phi$ $(\alpha \in H(\tau_M))$. Let γ be the function on M defined by

$$\gamma(m) = \tau_M(m) \qquad (m \in K \cap M) ,$$

$\gamma(m) = 0$ for m outside $K \cap M$. Then $\gamma \in H(\tau_M)$. Put

$$\gamma_a(m) = \gamma(ma) \qquad (m \in M, \ a \in A) .$$

Then $\gamma_a \in H(\tau_M)$, $\gamma_{a_1 a_2} = \gamma_{a_1} * \gamma_{a_2}$ and $\gamma_a * \psi = \rho(a)\psi$ $(\psi \in \Phi)$, where $\rho(a)$ is the right-translation over a $(a \in A)$. Therefore A operates on Φ by right translations.

If $\Phi \neq \{0\}$, we can choose $\phi_o \neq 0$ in Φ and a quasi-character χ of A such that

$$\rho(a)\phi_o = \chi(a)\phi_o \qquad (a \in A) ,$$

i.e. $\phi_o(ma) = \chi(a)\phi_o(m)$ $(a \in A, m \in M)$. Now $\phi \underset{P}{\approx} 0$, hence $\alpha * \phi \underset{P}{\approx} 0$ for all $\alpha \in H(\tau_M)$. In particular $\phi_o \underset{P}{\approx} 0$. This implies $\phi_o(ma) = 0$ for $a \in A^+(c)$ for some $c > 0$ (depending on m). Choose m such that $\phi_o(m) \neq 0$. Then $\phi_o(ma) = \phi_o(m)\chi(a) \neq 0$ for all $a \in A$. Contradiction. Therefore $\Phi = \{0\}$ and hence $\phi = \gamma * \phi = 0$. This completes the proof of the lemma.

Lemma 11. Let $K(\tau)$ denote the kernel of τ. There exists a finite set of elements $a_1, \ldots, a_r \in A^+$ with the following property. Let S denote the semi-group generated by $a_1, \ldots, a_r$. Then given $t > 0$, we can choose a finite subset B of A such that

$$A^+(t) \subset B.S.A \cap K(\tau) .$$

Proof. Let $\alpha_1, \ldots, \alpha_\ell$ be the elements of $\Sigma^o = \Sigma^o(P/A)$. We can choose $a_1, \ldots, a_\ell \in A$ such that $\xi_{\alpha_i}(a_j) = 1$ $(i \neq j)$ and $|\xi_{\alpha_i}(a_i)| > 1$. Let D^+ be the semi-group generated by $a_1, \ldots, a_\ell$. Let $t > 0$ be given. We claim: there is a compact subset $\omega \subset A$ such that

$$A^+(t) \subset \omega Z D^+$$

(as usual Z is the split component of G). Indeed, suppose $a \in A^+(t)$. Let p_i be the least non-negative integer such that

$$|\xi_{\alpha_i}(a_i)|^{p_i} \geq |\xi_{\alpha_i}(a)| \qquad (1 \leq i \leq \ell) .$$

Define $b \in A$ by $a = b\, a_1^{p_1} a_2^{p_2} \ldots a_\ell^{p_\ell}$. Then

$$|\xi_{\alpha_i}(b)| = |\xi_{\alpha_i}(a)|\,|\xi_{\alpha_i}(a_i)|^{-p_i} \leq 1 .$$

If $p_i = 0$, then $|\xi_{\alpha_i}(b)| = |\xi_{\alpha_i}(a)| \geq t$; if $p_i \geq 1$, then $|\xi_{\alpha_i}(a)| > |\xi_{\alpha_i}(a_i)|^{p_i - 1}$, hence $|\xi_{\alpha_i}(b)| > |\xi_{\alpha_i}(a_i)|^{-1}$. In any case

$$1 \geq |\xi_{\alpha_i}(b)| \geq \min(t, |\xi_{\alpha_i}(a_i)|^{-1}) .$$

Since $Z = \bigcap_{\alpha \in \Sigma^o} \ker \xi_\alpha$, we have $b \in \omega Z$ for some compact set $\omega \subset A$. Hence $A^+(t) \subset \omega Z D^+$.

Now $Z/Z \cap K(\tau)$ is a finitely generated abelian group. Let $z_1, \ldots, z_p$ be elements in Z which generate $Z \bmod Z \cap K(\tau)$. Define $F = \{a_1, \ldots, a_\ell, z_1, z_1^{-1}, \ldots, z_p, z_p^{-1}\} \subset A^+$. Let S be the semi-group generated by F. We have

$$A^+(t) \subset \omega . S . Z \cap K(\tau) \subset \omega . S . A \cap K(\tau) ,$$

and since $\omega . A \cap K(\tau)$ is compact and $A \cap K(\tau)$ is open in A, there is a finite set B such that

$$A^+(t) \subset B . S . A \cap K(\tau) .$$

This proves the lemma.

Lemma 12. Let g be a function $G \longrightarrow \mathrm{End}(V)$ satisfying

$$g(k_1 x k_2) = \tau(k_1) g(x) \tau(k_2) \qquad (k_1, k_2 \in K, x \in G) .$$

Let $\beta \in H(\tau)$. Then

$$(\beta * g)_{(P)} \underset{P}{\approx} \beta^{(P)} * g_{(P)} .$$

Proof. Let ω be a compact subset of M. For $m_o \in \omega$, $a_o \in A$ we have

$$(\beta*g)(m_o a_o) = \int_G \beta(x^{-1})g(xm_o a_o)dx = \int_K\int_N\int_M \beta(m^{-1}n^{-1}k^{-1})g(knmm_o a_o)dkdndm$$

$$= \int_N\int_M \beta(m^{-1}n^{-1})g(mm_o a_o n^{(mm_o a_o)^{-1}})dndm \ .$$

Since m, m_o and n remain bounded, we have $n^{(mm_o a_o)^{-1}} \in K(\tau)$ if $\gamma_P(a_o) \geq c$ for some large $c > 0$. This being the case, we obtain

$$(\beta*g)(m_o a_o) = \int_N\int_M \beta(m^{-1}n^{-1})g(mm_o a_o)dndm \ .$$

On the other hand,

$$\begin{aligned}(\beta*g)_{(P)}(m_o a_o) &= \delta_P(m_o a_o)^{1/2}(\beta*g)(m_o a_o)\\ &= \delta_P(m_o a_o)^{1/2}\int_N\int_M \beta(m^{-1}n^{-1})\delta_P(m^{-1})^{1/2}\delta_P(m)^{1/2}g(mm_o a_o)dndm\\ &= \beta^{(P)} * g_{(P)}(m_o a_o)\end{aligned}$$

if $\gamma_P(a_o) \geq c$ and $m_o \in \omega$. The assertion of the lemma is now obvious.

Lemma 13 (based on Conjecture III). __Fix__ $f \in \mathcal{A}(G, \tau)$. __Let__ F __be the space of all functions on__ M __of the form__ $\alpha * f_{(P)}$ $(\alpha \in H(\tau_M))$ __and let__ F_o __be the subspace of all__ $\phi \in F$ __with__ $\phi \underset{P}{\approx} 0$. __Then__ $\dim F/F_o < +\infty$.

__Proof.__ Put $F_G = H(\tau) * f$. Since $f \in \mathcal{A}(G, \tau)$, $\dim F_G < +\infty$. We have a representation σ of $H(\tau)$ on F_G given by

$$\beta * \phi = \sigma(\beta)\phi \qquad\qquad (\beta \in H(\tau),\ \phi \in F_G) \ .$$

Hence $\dim H(\tau)/\mathrm{Ker}\ \sigma < +\infty$.

If $\beta \in \ker \sigma$, then $\beta * f = 0$. So by Lemma 12, $\beta^{(P)} * f_{(P)} \in F_o$. Let as before μ_P denote the homomorphism $\beta \longmapsto \beta^{(P)}$ of $H(\tau)$ into $H(\tau_M)$. Then we have $\mu_P(\ker \sigma) * f_{(P)} \subset F_o$. Notice that F_o is stable under convolution on the left by elements of $H(\tau_M)$. Conjecture III gives

$$H(\tau_M) = \sum_{1\leq i\leq r} a_i * \mu_P(H(\tau)) .$$

Since $H(\tau) = \sum_{1\leq j\leq s} \mathbb{C}\beta_j + \ker \sigma$ for certain $\beta_j \in H(\tau)$ $(1 \leq j \leq s)$, we get

$$H(\tau_M) = \sum_{i,j} \mathbb{C}(a_i * \mu_P(\beta_j)) + \sum_i a_i * \mu_P(\ker \sigma) .$$

Hence

$$F = H(\tau_M) * f_{(P)} \subset \sum_{i,j} \mathbb{C}(a_i * \beta_j^{(P)} * f_{(P)}) + F_o .$$

So $\dim F/F_o < +\infty$ and the lemma follows.

Let $f \in \mathcal{A}^o(G, \tau)$ and let us adopt the notations of Lemma 13. Assume $f_{(P)} \notin F_o$. Since $f_{(P)} \in F$, we can choose a base

$$\phi_1 = f_{(P)}, \phi_2, \ldots, \phi_p$$

of F mod F_o. Since F and F_o are both stable under convolution on the left by elements of $H(\tau_M)$, we obtain a representation σ of $H(\tau_M)$ on F mod F_o given by $a * \phi_i = \sum_j \phi_j \sigma_{ji}(a)$ mod F_o, $\sigma(a) = (\sigma_{ji}(a))$. Put

$$\Phi = \begin{pmatrix} \phi_1 \\ \vdots \\ \phi_p \end{pmatrix} .$$

Then, in an obvious notation,

$$a * \Phi \underset{P}{\approx} \Phi\sigma(a) \qquad (a \in H(\tau_M)) .$$

As in the proof of Lemma 10, let $\gamma \in H(\tau_M)$ be defined by

$$\gamma(m) = \tau_M(m) \text{ for } m \in K \cap M, \text{ Supp } \gamma \subset K \cap M .$$

For $a \in A$, put $\gamma_a(m) = \gamma(am)$ $(m \in M)$. Then $\gamma_{a_1 a_2} = \gamma_{a_1} * \gamma_{a_2}$. Denote by χ the representation of A on F mod F_o given by $a \longmapsto \sigma(\gamma_a)$. So $\gamma_a * \Phi \underset{P}{\approx} \Phi\chi(a)$, or

$$\rho(a)\Phi \underset{P}{\approx} \Phi\chi(a) \qquad (a \in A)$$

where $\rho(a)$ is the right-translation over $a \in A$. Put

$$\Psi_a = \rho(a)\Phi - \Phi\chi(a) \qquad (a \in A) \quad .$$

Lemma 14. Fix a compact set ω in M and $t > 0$. Then we can choose $c > 0$ such that $\Psi_a = 0$ on $\omega A^+(c)$ for all $a \in A^+(t)$.

Proof. Fix ω and $t > 0$. By Lemma 11 we can choose $a_1, \ldots, a_r \in A^+$ such that $A^+(t) \subset B.S.A \cap K(\tau)$ for some finite subset $B \subset A$, where S is the semi-group generated by $a_1, \ldots, a_r$. Choose $c > 0$ so large that

$$\Psi_b = \Psi_{ba_i} = 0 \text{ on } \omega A^+(c) \qquad (b \in B,\ 1 \leq i \leq r) \quad .$$

Put $a = a_1^{n_1} a_2^{n_2} \ldots a_r^{n_r}$, $n = \sum_{1 \leq i \leq r} n_i$. We claim that $\Psi_{ba} = 0$ on $\omega A^+(c)$ for all $b \in B$. We prove this by induction on n. Assume $n \geq 2$, $n_1 \geq 1$. Put $a_o = a_1^{n_1 - 1} a_2^{n_2} \ldots a_r^{n_r}$. Then $a = a_1 a_o$. By the induction hypothesis $\Psi_{ba_o} = 0$ on $\omega A^+(c)$. So $\Phi(mba_o) - \Phi(m)\chi(ba_o) = 0$ for $m \in \omega A^+(c)$. If $m \in \omega A^+(c)$, so is ma_1. Hence

$$\Phi(ma_1 ba_o) - \Phi(ma_1)\chi(ba_o) = 0 \qquad (m \in \omega A^+(c)) \quad .$$

Now $\Phi(ma_1) = \Phi(m)\chi(a_1)$ for $m \in \omega A^+(c)$ and it follows

$$\Phi(mba_1 a_o) - \Phi(m)\chi(ba_1 a_o) = 0 \qquad (m \in \omega A^+(c)) \quad .$$

This implies $\Psi_{ba} = 0$ on $\omega A^+(c)$. We have proved: $\Psi_a = 0$ on $\omega A^+(c)$ for all $a \in B.S$. Since $\Psi_{aa'} = \Psi_a$ for $a \in A$, $a' \in A \cap K(\tau)$, we have $\Psi_a = 0$ on $\omega A^+(c)$ for all $a \in B.S.A \cap K(\tau)$. Therefore, a fortiori, $\Psi_a = 0$ on $\omega A^+(c)$

for all $a \in A^+(t)$. This proves the lemma.

§3. Proof of Theorem 7.

Fix $f \in \mathcal{A}(G, \tau)$. We may assume $f_{(P)} \in {}^e F_o$, otherwise we can choose $f_P = 0$. We have to find $f_P \in \mathcal{A}(M, \tau_M)$ such that $f_{(P)} - f_P \underset{P}{\approx} 0$. The uniqueness of f_P follows from Lemma 10.

Let ω be a compact subset of M. By Lemma 14 we can choose $c > 0$ such that $\Phi(ma) = \Phi(m)\chi(a)$ for $a \in A^+$ and $m \in \omega A^+(c)$. Put $c_o = \max(1, c)$. Then

$$\Phi(ma) = \Phi(m)\chi(a) \quad \text{for } a \in A^+(c_o),\ m \in \omega A^+(c_o) .$$

Consequently

$$\Phi(ma_1 a_2) = \Phi(ma_1)\chi(a_2) \qquad (m \in \omega;\ a_1,\ a_2 \in A^+(c_o)) .$$

By interchanging the role of a_1 and a_2 we get

$$\Phi(ma_1)\chi(a_1)^{-1} = \Phi(ma_2)\chi(a_2)^{-1} \qquad (m \in \omega;\ a_1,\ a_2 \in A^+(c_o)) .$$

We write $a \underset{P}{\longrightarrow} \infty$ if $\gamma_P(a) \longrightarrow \infty$. Define

$$\Theta(m) = \lim_{a \underset{P}{\to} \infty} \Phi(ma)\chi(a)^{-1} \qquad (m \in M) .$$

Θ is locally constant. Furthermore

$$\Theta(ma_o) = \lim_{a \underset{P}{\to} \infty} \Phi(ma_o a)\chi(a)^{-1} = \lim_{a \underset{P}{\to} \infty} \Phi(ma)\chi(a)^{-1}\chi(a_o) = \Theta(m)\chi(a_o)$$

for $m \in M$, $a_o \in A$. For $m \in \omega$, $a \in A^+(c_o)$ we have

$$\Theta(m) = \Phi(ma)\chi(a)^{-1} .$$

So, given ω, there exists $c_o > 0$ such that

$$\Theta(m) = \Phi(m) \quad \text{for } m \in \omega A^+(c_o) .$$

Hence $\Theta \underset{P}{\approx} \Phi$. Now

$$\Theta = \begin{pmatrix} \theta_1 \\ \vdots \\ \theta_p \end{pmatrix},$$

so $\phi_i \underset{P}{\approx} \theta_i$ $(1 \leq i \leq p)$. In particular $f_{(P)} \underset{P}{\approx} \theta_1$. We verify that θ_1 belongs to $\mathcal{A}(M, \tau_M)$. Write θ for θ_1. We have

$$\theta(kmk') = \tau_M(k)\theta(m)\tau_M(k') \qquad (k,\ k' \in K \cap M,\ m \in M)$$

since Φ has this property. To prove that θ is left $H(\tau_M)$-finite, put

$$\Theta_\alpha = \alpha * \Theta - \Theta\sigma(\alpha) \qquad (\alpha \in H(\tau_M)) .$$

Since $\alpha * \Phi - \Phi\sigma(\alpha) \underset{P}{\approx} 0$, we have also $\Theta_\alpha \underset{P}{\approx} 0$. We know

$$\rho(a)\Theta = \Theta\chi(a) \qquad (a \in A) .$$

Since γ_a is in the center of $H(\tau_M)$ (a is in the center of M), we get

$$\rho(a)\Theta_\alpha = \Theta_\alpha\chi(a) \qquad (a \in A,\ \alpha \in H(\tau_M)) .$$

It follows that both m and ma belong to Supp Θ_α for all $a \in A$.* So Supp $\Theta_\alpha \neq \emptyset$ contradicts $\Theta_\alpha \underset{P}{\approx} 0$. Hence $\alpha * \Theta = \Theta\sigma(\alpha)$ for all $\alpha \in H(\tau_M)$. So $\alpha * \theta = \sum_{1 \leq j \leq p} \sigma_{j1}(\alpha)\theta_j$ for all $\alpha \in H(\tau_M)$. This shows that θ is left $H(\tau_M)$-finite and hence $\theta \in \mathcal{A}(M, \tau_M)$. To complete the proof, take $f_P = \theta$.

§4. Proof of Theorem 8.

We prove the following version of Theorem 8, based again on Conjecture III.

---

*This can be seen by choosing one ϕ_i such that $\rho(a)\phi_i = \chi_o(a)\phi_i (\text{mod } F_o)$ for some quasi-character χ_o of A.

Theorem 8'. Fix a mincuspidal pair (P_o, A_o) as in the theorem of Bruhat and Tits. Let $f \in \mathcal{A}(G, \tau)$ and suppose $f_P = 0$ for all $(P, A) \succ (P_o, A_o)$, $P \neq G$. Then f is a supercusp form.

Proof. We first show that Supp f is compact mod Z. Let $P_o = M_o N_o$ be the Levi-decomposition of (P_o, A_o). By Theorem 5 we have $G = KCA_o^+K$, C being a finite subset of M_o. It suffices to show that for all $c \in C$, $f(ca) = 0$ unless

$$\max_{a \in \Sigma^o(P_o/A_o)} |\xi_a(a)| \leq t$$

for some $t > 0$. Assume that no such a $t > 0$ exists. Then we can choose $c \in C$, $a_k \in A_o^+$ and $a_k \in \Sigma^o = \Sigma^o(P_o/A_o)$ such that $f(ca_k) \neq 0$ and $|\xi_{a_k}(a_k)| \geq k$ $(k = 1, 2, \ldots)$.

Let F be the set of roots $a \in \Sigma^o$ such that $\sup_k |\xi_a(a_k)| < +\infty$. So we have

$$|\xi_a(a_k)| \longrightarrow \infty \text{ for } a \in {}^cF, \quad \sup_k |\xi_a(a_k)| < +\infty \text{ for } a \in F .$$

Let $(P, A) = (P_o, A_o)_F$. Let $P = MN$ be the Levi-decomposition of the pair (P, A). Since ${}^cF \neq \emptyset$, we have $P \neq G$. Observe that $M_o \subset M$. Write $a_k = m_k a_k'$, m_k remaining bounded in M, $a_k' \in A$ and $|\xi_a(a_k')| \longrightarrow \infty$ for $a \in {}^cF \cong \Sigma^o(P/A)$. Put $\omega = \{cm_k\}$ $(c \in C, k = 1, 2, \ldots)$. Then ω is a relatively compact subset of M. Furthermore $\delta_P(ca_k)^{1/2} f(ca_k) = f_{(P)}(cm_k a_k') = f_P(cm_k a_k')$ for large k since $\gamma_P(a_k') \longrightarrow \infty$ (Theorem 7). Now $f_P = 0$ since $(P, A) \succ (P_o, A_o)$ and $P \neq G$. Hence $f(ca_k) = 0$ for k large. Contradiction! The remaining part of the theorem follows from the next lemma.

Lemma 15 (based on Conjecture III). If $f \in \mathcal{A}(G, \tau)$ has compact support mod Z, then f is a supercusp form.

Proof. We use induction on ssrk G. If ssrk G = 0, then G/Z is compact. Every continuous function on G is a supercusp form.

Assume ssrk G > 0. Let (P, A) be a cuspidal pair, $P \neq G$, P = MN the corresponding Levi-decomposition. We have ssrk M < ssrk G. It is clear that $f^{(P)}$ can be defined. For all $\alpha \in H(\tau)$,

$$(\alpha * f)^{(P)} = \alpha^{(P)} * f^{(P)} = \mu_P(\alpha) * f^{(P)} .$$

So $f^{(P)}$ is left $\mu_P(H(\tau))$-finite. Applying Conjecture III it follows easily that $f^{(P)} \in \mathcal{A}(M, \tau_M)$. Put $\phi = f^{(P)}$ and $\Phi = H(\tau_M) * \phi$. Then $\dim \Phi < +\infty$.

As in the proof of Lemma 10, we have a representation ρ of A on Φ. If $\Phi \neq \{0\}$, we can choose $\phi_o \neq 0$ in Φ and a quasi-character χ_o of A such that $\rho(a)\phi_o = \chi_o(a)\phi_o$ $(a \in A)$. So Supp ϕ_o = Supp $\phi_o \cdot A$. Since A/Z is not compact, this yields a contradiction. Therefore $\Phi = \{0\}$ and hence $f^{(P)} = 0$. We have

$$\int_N f(xn)dn = \int_N f(k_o m_o n)dn = \tau(k_o) \int_N f(m_o n)dn$$
$$= \tau(k_o)\delta_P(m_o)^{-1/2} f^{(P)}(m_o) = 0$$

for all $x \in G$, $x = k_o m_o n_o$. Hence $f^P = 0$ for all $P \neq G$. This implies that f is a supercusp form and the lemma follows.

We conclude this Part with two theorems which may be considered as 'problems'. Compare them with the similar results in the real case (cf. Harish-Chandra, Colloquium Lectures, University of Oregon, A.M.S. 1969, §9).

Given $f \in \mathcal{A}(G, \tau)$, we write $f_P \sim 0$ if

$$\int_M (\phi(m), f_P(m))dm = 0 \text{ for all } \phi \in {}^oC_c(M, \tau_M) .$$

Here the scalar product is taken in V_P.

Theorem A. Let $f \in \mathcal{A}(G, \tau)$ and suppose $f_P \sim 0$ for all P (including P = G). Then f = 0.

Theorem B. *Let* $f \in \mathcal{A}(G, \tau)$, $f \neq 0$. *Let* $P = MN$ *be a parabolic subgroup of* G *such that* $f_P \neq 0$ *and* P *minimal with respect to this property. Then* f_P *is a supercusp form in* $\mathcal{A}(M, \tau_M)$.

Part V. The behavior of the characters of the supercuspidal representations on the regular set.

In this Part we begin a closer study of the characters of the irreducible unitary representations of a reductive $\mathfrak{p}$-adic group G. In view of the 'philosophy of cusp forms' and the results of Part I, we confine ourselves to supercuspidal representations. We shall prove that their characters actually are locally summable functions on G, which are locally constant on G', the set of regular elements of G. Unfortunately we can prove the local summability only in case char $\Omega = 0$ (cf. next Parts). Jacquet and Langlands have established the above statement for $G = GL(2)$ without any restriction on char Ω (cf. H. Jacquet - R. P. Langlands, Automorphic forms on GL(2), Springer Lecture Notes 114 (1970), §7). We were not able to generalize their method for the case of positive characteristic.

§1. Two fundamental theorems.

We start with a theorem which is easily proved in the case of $\mathfrak{p}$-adic groups. It is however still true under more general circumstances. In this paragraph G is a reductive $\mathfrak{p}$-adic group.

Theorem 9. Let $\omega \in \mathcal{E}_2(G)$. Fix $\pi \in \omega$ and denote by $\mathcal{H}_\pi$ the space of π. Then for any $f \in C_c^\infty(G)$

$$\operatorname{tr} \pi(f)\,(\phi, \psi) = d(\omega) \int_{G/Z} dx^* \int_G f(y)\,(\phi, \pi(y^x)\psi)\,dy$$
$$= d(\omega) \int_{G/Z} (\phi, \pi(x)\pi(f)\pi(x^{-1})\psi)\,dx^*$$

for all $\phi, \psi \in \mathcal{H}_\pi$.

Proof. By Theorem 3, $Q = \pi(f)$ is of finite rank. There is an open compact subgroup K_o of G such that

$$f(kxk') = f(x) \qquad (x \in G;\ k, k' \in K_o)\ .$$

Put $E = \int_{K_o} \pi(k)dk$, where the Haar measure dk of K_o is normalized such that the total measure of K_o is one. Then $Q = EQE$. Put $Q^x = \pi(x)Q\pi(x^{-1})$.

Since $\dim E(\mathcal{H}_\pi) < +\infty$ (cf. corollary of Theorem 2), we can choose an orthonormal base $(\phi_i)_{i\in J}$ of $E(\mathcal{H}_\pi)$ with Card $J < +\infty$. Put

$$Q_{ij} = (\phi_i, Q\phi_j) \qquad (i, j \in J) .$$

We have

$$(\phi, Q^x\psi) = (\pi(x^{-1})\phi, Q\pi(x^{-1})\psi) = \sum_{i,j\in J} (\pi(x^{-1})\phi, \phi_i)\cdot Q_{ij}\cdot(\phi_j, \pi(x^{-1})\psi)$$

$$= \sum_{i,j\in J} Q_{ij}(\phi, \pi(x)\phi_i)\overline{(\psi, \pi(x)\phi_j)} \qquad (\phi, \psi \in \mathcal{H}_\pi) .$$

So $x \longmapsto (\phi, Q^x\psi)$ is integrable mod Z and

$$\int_{G/Z} (\phi, Q^x\psi)dx^* = \sum_{i,j\in J} Q_{ij} d(\omega)^{-1} (\phi_j, \phi_i)(\phi, \psi)$$

$$= d(\omega)^{-1}(\phi, \psi)\sum_{i\in J} Q_{ii} = d(\omega)^{-1}(\phi, \psi) \operatorname{tr} Q \qquad (\phi, \psi \in \mathcal{H}_\pi) .$$

This proves the theorem.

Let $\omega \in {}^o\mathcal{E}(G)$ and fix $\pi \in \omega$. Let ξ be a K-finite unit vector in $\mathcal{H}_\pi$, the space of π. Then $\theta(x) = (\xi, \pi(x)\xi)$ $(x \in G)$ satisfies:

$$\operatorname{tr} \pi(f) = d(\omega) \int_{G/Z} dx^* \int_G f(y)\theta(y^x)dy$$

for all $f \in C_c^\infty(G)$.

Now, to obtain the desired result, the idea is to interchange the integrals (roughly speaking).

For the next theorem we introduce some notation. Fix, once for all in this Part, a mincuspidal pair (P_o, A_o) and choose a corresponding open compact subgroup K as in the theorem of Bruhat and Tits. Let $A \subset A_o$ be a split component of some parabolic subgroup. Let $\mathcal{P} = \mathcal{P}(A)$ be the set of all parabolic subgroups P which have A as a split component. Furthermore denote by $\mathcal{P}' = \mathcal{P}'(A)$ the set of all cuspidal pairs (P', A') such that $A' \subset A$, $P' \neq G$. Both $\mathcal{P}$ and $\mathcal{P}'$ are finite sets.

Now fix a cuspidal pair $(P', A') \in \mathcal{P}'$ and let $P' = M'N'$ be the corresponding Levi-decomposition. Let $\overline{N'}$ be the opposite of N' (corresponding to the negative roots w.r.t. (P', A')). It is known that the map $(\overline{n}', m', n') \longmapsto \overline{n}'m'n'$ is an Ω-isomorphism of algebraic varieties of $\overline{N'} \times M' \times N'$ onto an open subset of G (cf. Borel-Tits, 4.2). In particular, $(\overline{N'} \cap K)(M' \cap K)(N' \cap K)$ is an open compact neighborhood of 1 in G.

Theorem 10. Given any neighborhood K_1 of 1 in G, there exists an open compact subgroup K_o of G contained in K_1 and satisfying the following condition. For any compact set $C \subset G$, we can choose another compact set $\omega \subset G$ with the following property. Let f be any continuous function such that

(i) $f(xa) = f(x)$ $(x \in G,\ a \in A)$, $\operatorname{Supp} f \subset CA$.

(ii) If $(P', A') \in \mathcal{P}'$, $P' = M'N'$, then

$$\int_{N'} f(xn)dn = 0 \qquad (x \in G) .$$

Then $\int_{K_o} f(xk)dk = 0$ unless $x \in \omega Z$ $(x \in G)$.

Proof. We may assume $K_1 \subset (\overline{N'} \cap K)(M' \cap K)(N' \cap K)$ for all cuspidal pairs $(P', A') \in \mathcal{P}'$, $P' = M'N'$.

The proof of the theorem is by contradiction (unfortunately). Choose an open compact subgroup $K_o \subset K_1$. We can choose a non-empty compact set C and a sequence of functions f_i satisfying (i) and (ii), as well as a sequence of points $x_i \in G$, such that $x_i \longrightarrow \infty \pmod Z$ and

$$\int_{K_o} f_i(x_i k)dk \neq 0 .$$

In particular there exist for all i elements $k_i \in K_o$, $y_i \in C$ and $a_i \in A$

such that $x_i k_i = y_i a_i$, $x_i k_i \in \operatorname{Supp} f_i$. Since k_i and y_i remain bounded, we have $a_i \longrightarrow \infty \pmod{Z}$.

Now let k be arbitrary in K_o. Write

$$x_i k = x_i k_i (k_i^{-1} k) = y_i a_i k_i'$$

where $k_i' = k_i^{-1} k$. Notice that $k_i' \in K_o$. Let $P \in \mathcal{P}(A)$ and put $A^+(P)$ = set of all $a \in A$ such that

$$|\xi_\alpha(a)| \geq 1 \text{ for all } \alpha \in \Sigma^o(P/A) \ .$$

Obviously $A = \bigcup_{P \in \mathcal{P}(A)} A^+(P)$. Since $\mathcal{P}(A)$ is a finite set, we may assume (by choosing a subsequence) that $a_i \in A^+(P)$ for some fixed $P \in \mathcal{P}(A)$ and all i. Since $a_i \longrightarrow \infty \pmod{Z}$, we have

$$\max_{\alpha \in \Sigma^o(P/A)} |\xi_\alpha(a_i)| \longrightarrow \infty \ .$$

Divide the roots in two classes. Let $F \subset \Sigma^o(P/A)$ be the set of all α satisfying $\sup_i |\xi_\alpha(a_i)| < +\infty$. So $|\xi_\alpha(a_i)| \longrightarrow \infty$ if $\alpha \in {}^cF$. Put $(P', A') = (P, A)_F$. Then, since ${}^cF \neq \emptyset$, $(P', A') \in \mathcal{P}'$. Let $P' = M'N'$ be the Levi-decomposition of the pair (P', A'). We had $x_i k = y_i a_i k_i'$ with $k_i' \in K_o \subset K_1 \subset (\overline{N'} \cap K)(M' \cap K)(N' \cap K)$. Write

$$k_i' = \overline{n}_i m_i n_i \quad (\overline{n}_i \in \overline{N'} \cap K, \ m_i \in M' \cap K, \ n_i \in N' \cap K)$$

and

$$x_i k = y_i a_i k_i' = y_i a_i \overline{n}_i m_i n_i = y_i (\overline{n}_i m_i)^{a_i} n_i^{a_i} a_i \ .$$

Put $z_i = y_i (\overline{n}_i m_i)^{a_i}$. Then

$$x_i k = z_i n_i^{a_i} . a_i \ .$$

We claim that z_i remains bounded. Since $\sup_i |\xi_\alpha(a_i)| < +\infty$ for $\alpha \in F$, we may write $a_i = b_i a_i'$ where $b_i \in A$, b_i remaining bounded mod Z, $a_i' \in A'$,

$|\xi_\alpha(a_i')| \longrightarrow \infty$ for $\alpha \in {}^cF$. Now

$$(\bar{n}_i m_i)^{a_i} = \bar{n}_i^{\,b_i a_i'} m^{b_i} \in (\bar{N}' \cap K)^{b_i a_i'} (M' \cap K)^{b_i} .$$

$(N' \cap K)^{b_i a_i'}$ remains bounded when i varies, since $\bar{N}'$ corresponds to the negative roots w.r.t. (P', A'); $(M' \cap K)^{b_i}$ is bounded since b_i is bounded mod Z. Hence there exists a compact set C_o, independent of $k \in K_o$ such that $z_i \in C_o$. Put $\omega' = (KC_o^{-1}CA) \cap N'$. This set is open and compact in N'. Indeed, let $\omega_{P'} = (KC_o^{-1}C) \cap P'$. This set is compact. Notice that $\omega' \subset \omega_{P'} A$. So if $n_o \in \omega'$, we have

$$n_o = mna = man^{a^{-1}} \qquad (mn \in \omega_{P'},\ m \in M',\ n \in N',\ a \in A)$$

and thus $ma = 1$, so $n_o = n^m$. Both m and n remain bounded, so n_o too. Now choose i so large $(i > i_o)$ that

$$(\omega')^{a_i^{-1}} = (\omega')^{b_i^{-1} a_i'^{-1}} \subset N' \cap K_o .$$

Normalize the Haar measure dn on N' in such a way that $\int_{K_o \cap N'} dn = 1$. Define

$$J_i(k) = \int_{K_o \cap N'} f_i(x_i kn)dn \qquad (k \in K_o) .$$

We claim that $J_i(k) = 0$ $(k \in K_o)$ for $i > i_o$. By assumption we have

$$\int_{N'} f_i(z_i n)dn = 0 .$$

Hence, moreover,

$$\int_{N'} f_i(z_i n^{a_i})dn = \delta_P(a_i^{-1}) \int_{N'} f_i(z_i n)dn = 0 .$$

If $f_i(z_i n) \neq 0$, then $z_i n \in CA$, so $n \in C_o^{-1}CA \subset \omega'$. Hence $f_i(z_i n^{a_i}) \neq 0$ implies $n \in (\omega')^{a_i^{-1}} \subset N' \cap K_o$ $(i > i_o)$. We compute

$$J_i(k) = \int_{K_o \cap N'} f_i(x_i kn)dn = \int_{K_o \cap N'} f_i(z_i(n_i n)^{a_i})dn .$$

If $n_i \in {}^c(K_o \cap N')$, then also $n_i n \in {}^c(K_o \cap N')$ for $n \in K_o \cap N'$, hence $J_i(k) = 0$ $(i > i_o)$. Now assume $n_i \in K_o \cap N'$. Then

$$J_i(k) = \int_{K_o \cap N'} f_i(z_i n^{a_i})dn = \int_{N'} f_i(z_i n^{a_i})dn = 0 .$$

Therefore $J_i(k) = 0$ $(k \in K_o)$ for $i > i_o$. But then

$$\int_{K_o} f_i(x_i k)dk = \int_{K_o/K_o \cap N'} dk \int_{K_o \cap N'} f_i(x_i kn)dn =$$

$$= \int_{K_o/K_o \cap N'} J_i(k)dk = 0 \text{ for } i > i_o .$$

This contradicts our assumption and completes the proof.

§2. $\mathfrak{p}$-Adic manifolds and distributions.

We introduce some general facts concerning $\mathfrak{p}$-adic manifolds and distributions. We refer to: Harish-Chandra, Invariant distributions on Lie algebras, Amer. J. Math. 86 (1964), 271-309, §§2, 3, 5.

Definition. A $\mathfrak{p}$-adic (analytic) manifold of dimension n is a Hausdorff space M together with a family Φ of mappings satisfying:

(i) each $\phi \in \Phi$ is a homeomorphism of an open set U_ϕ in M and $V_\phi = \phi(U_\phi)$, an open set in Ω^n;

(ii) if $\phi_1, \phi_2 \in \Phi$, then the homeomorphism of $\phi_1(U_{\phi_1} \cap U_{\phi_2})$ onto $\phi_2(U_{\phi_1} \cap U_{\phi_2})$ defined by $\phi_2 \cdot \phi_1^{-1}$ is analytic;

(iii) $(U_\phi)_{\phi \in \Phi}$ is a covering for M;

(iv) the family Φ is maximal with respect to the properties (i), (ii), and (iii).

Definition. A volume element dv on M is a positive measure on M

such that, given any point $x_o \in M$ and a coordinate system $t = (t_1, \ldots, t_n)$ around x_o, one has $dv = c\,dt$ $(c > 0)$, where $dt = dt_1 \ldots dt_n$, in a sufficiently small neighborhood of x_o.

Theorem 11. Let ψ be an analytic mapping of a $\mathfrak{p}$-adic manifold M onto a $\mathfrak{p}$-adic manifold N which is everywhere submersive. Let dv_M and dv_N be two volume elements on M and N respectively. Then for every $\alpha \in C_c^\infty(M)$ there exists exactly one function $f_\alpha \in C_c^\infty(N)$ such that for all $F \in C_c^\infty(N)$

$$\int \alpha(F \circ \psi)dv_M = \int f_\alpha F\, dv_N .$$

Moreover $\alpha \longmapsto f_\alpha$ is a surjective linear mapping of $C_c^\infty(M)$ onto $C_c^\infty(N)$. Furthermore $\operatorname{Supp} f_\alpha \subset \psi(\operatorname{Supp} \alpha)$.

The proof is exactly the same as for real manifolds (cf. the above reference, Theorem 1).

Corollary. Let F be a measurable function on N. Then F is locally summable on N iff $F \circ \psi$ is locally summable on M. If this is so, then

$$\int \alpha(F \circ \psi)dv_M = \int f_\alpha F\, dv_N \qquad (\alpha \in C_c^\infty(M)) .$$

Definition. A distribution T on a $\mathfrak{p}$-adic manifold M is a linear function on $C_c^\infty(M)$.

In the notations of Theorem 11, let S be a distribution on N. Define $\tau_S(\alpha) = T(f_\alpha)$ $(\alpha \in C_c^\infty(M))$. Then τ_S is obviously a distribution on M.

Let U be an open subspace of M. Let T be a distribution on M. We say that T is a function on U if there exists a locally summable function (w.r.t. dv_M) F on U such that

$$T(\alpha) = \int \alpha F\, dv_M \qquad (\alpha \in C_c^\infty(U)) .$$

In this case we write: $T = F$ on U.

Now let, more generally, M be a locally compact zero-dimensional Hausdorff space. Denote by $C_c^\infty(M)$ the space of locally constant complex-valued functions with compact support on M. Distributions on M will be again linear functions on $C_c^\infty(M)$. Let V be any complex vector space.

Lemma 16. Denote by $C_c^\infty(M, V)$ the space of all locally constant functions with compact support from M to V. Then $C_c^\infty(M, V) = C_c^\infty(M) \otimes V$.

Corollary. $C_c^\infty(M_1 \times M_2) = C_c^\infty(M_1) \otimes C_c^\infty(M_2)$, where M_1 and M_2 are locally compact zero-dimensional Hausdorff spaces.

Let G be a locally compact zero-dimensional group, $d_\ell x$ a left Haar measure on G, $d_r x$ a right Haar measure and $d_r x = \delta(x) d_\ell x$. Assume that G operates on M. This means that for every $g \in G$ there is given a homeomorphism $x \longmapsto x^g$ of M such that $x^1 = x$ and $(x^{g_2})^{g_1} = x^{g_1 g_2}$ $(g_1, g_2 \in G,\ x \in M)$. Fix $g \in G$. Define

$$f^g(x) = f(x^{g^{-1}}) \qquad (f \in C_c^\infty(M),\ x \in M) .$$

Let T be a distribution on M. Define

$$T^g(f) = T(f^{g^{-1}}) \qquad (f \in C_c^\infty(M)) .$$

Then T^g is also a distribution. T is called invariant if $T = T^g$ for all $g \in G$.

Take $M = G$ and define $x^g = gx$ $(g, x \in G)$. T is called left-invariant if T is invariant w.r.t. this action of G on G.

Lemma 17. Let T be a left-invariant distribution on G. Then there exists a complex constant c such that

$$T(f) = c \int_G f(x) d_\ell x \qquad (f \in C_c^\infty(G)) .$$

Proof. Choose $\alpha \in C_c^\infty(G)$ such that $\int_G \alpha(y) d_\ell y = 1$. For any $f \in C_c^\infty(G)$, put $g(x) = \int_G \alpha(y)f(y^{-1}x)d_\ell y$. Then $T(g) = \int_G \alpha(y)T(f)d_\ell y = T(f)$ since T is left-invariant. On the other hand,

$$g(x) = \int_G \alpha(y^{-1})f(yx)d_\ell y^{-1} = \int_G \alpha(y^{-1})f(yx)d_r y =$$
$$= \int_G \alpha(xy^{-1})f(y)d_r y = \int_G (\rho(y^{-1})\alpha)(x)f(y)\delta(y)d_\ell y$$

where $\rho(y^{-1})\alpha(x) = \alpha(xy^{-1})$ $(x, y \in G)$. Hence

$$T(f) = T(g) = \int_G T(\rho(y^{-1})\alpha)f(y)\delta(y)d_\ell y = \int_G \beta(y)f(y)d_\ell y$$

where $\beta(y) = T(\rho(y^{-1})\alpha)\delta(y)$ $(y \in G)$. Applying again that T is left-invarinat, we get $\beta(y) = c$ for some complex constant c $(y \in G)$. This gives the lemma.

§3. Invariant distributions on the regular set.

We proceed with an algebraic lemma. Let A be a split torus in G. Denote by $\mathcal{Z}(A)$ the centralizer of A in G.

Lemma 18. A is a split component of a parabolic subgroup iff A is the maximal split torus in the center of $\mathcal{Z}(A)$.

Proof. We prove the 'if' part of the lemma. Denote by $X_\Omega(A)$ the (additive) group of all rational homomorphisms of $\underset{\sim}{A}$ into GL(1) defined over Ω. This is a free abelian group of rank $\ell = \dim \underset{\sim}{A}$. Take an order on the space $X_\Omega(A) \otimes_{\mathbb{Z}} \mathbb{R}$. Let N be the subgroup spanned by the elements in G corresponding to the positive roots of G w.r.t. A (in the order chosen above). So $\mathfrak{n} = \sum_{\alpha>0} \mathfrak{n}_\alpha$ where $\mathfrak{n}$ is the Lie algebra of N and $\mathfrak{n}_\alpha$ is the root space of the root α. Then $P = \mathcal{Z}(A)N$ is a parabolic subgroup of G (cf. Borel-Tits 4.15) with A as a split component. This proves the lemma.

By a Cartan subgroup of G, we mean a subgroup of the form $\Gamma = G \cap \underset{\sim}{\Gamma}$, where $\underset{\sim}{\Gamma}$ is a maximal Ω-torus in G. Such a subgroup always exists.

Let Γ be a Cartan subgroup of G and A its maximal split torus. Then A is a split component of a parabolic subgroup in G. Indeed, $\Gamma \subset \mathcal{Z}(A)$, hence Γ is also a Cartan subgroup of $\mathcal{Z}(A)$, so Γ contains the center of $\mathcal{Z}(A)$. Therefore A is the maximal split torus lying in the center of $\mathcal{Z}(A)$. The assertion follows now from Lemma 18.

Lemma 19. Let Γ be a Cartan subgroup of G. Fix $\gamma_o \in \Gamma$. Let Ξ be the centralizer of γ_o in G. Then there exists an open neighborhood ω_Γ of γ_o in Γ with the following property. Given any compact set ω_G in G, we can choose a compact set C^* in $G^* = G/\Xi$ such that

$$(\omega_\Gamma)^x \cap \omega_G Z = \emptyset$$

unless $x^* \in C^*$

Proof. The proof is in two steps.

(i) Reduction to the case of split Γ. Denote by L a finite separable field extension of Ω such that $\underset{\sim}{\Gamma}$ splits over L. Let L be endowed with the unique valuation which extends that of Ω. Γ_L is a split Cartan subgroup of G_L. Hence there exists a parabolic subgroup P of G_L with Γ_L as a split component. Let $P = N\Gamma_L$ be the corresponding Levi-decomposition. Furthermore there exists a compact (open) subgroup K of G_L such that $G_L = KN\Gamma_L$ (Bruhat-Tits). Obviously G is closed in G_L, $\Gamma = \Gamma_L \cap G$, $\Xi = \Xi_L \cap G$ since Ξ is defined over Ω (cf. A. Borel, Linear algebraic groups, Benjamin 1969, p. 224). So $G^* = G/\Xi$ may be regarded as a subset of $G_L^* = G_L/\Xi_L$. Now assume Lemma 19 to be true for the pair (G_L, Γ_L). Then Lemma 19 follows for (G, Γ), provided

(a) G^* is closed in G_L^*;

(b) the inclusion map $G^* \longrightarrow G_L^*$ is topological.

Therefore, let $\{x_r^*\}_{r \geq 1}$ be a sequence in G^* which converges in G_L^* to some point x^*. Then we have to show that $x^* \in G^*$ and x_r^* converges to x^* in G^*.

Select points $x_r \in G$ and $x \in G_L$ lying in the cosets x_r^* and x^* respectively. Then obviously $x_r\gamma_o x_r^{-1} \longrightarrow x\gamma_o x^{-1}$ $(r \longrightarrow \infty)$ in G_L and therefore $x_r\gamma_o x_r^{-1}$ remains in a compact subset of G_L. But this implies that x_r^* are all contained in a compact subset of G^* (cf. A. Borel, previous reference, p. 224-225) and from this (a) and (b) follows immediately.

(ii) Proof of the lemma in the split case. We assume Γ split (over Ω). As before, $G = KP = KN\Gamma$. Making use of this decomposition, it is easily seen that we are reduced to prove the existence of ω_Γ and in addition: there exists a compact subset $C^* \subset G^*$ such that, if $n\gamma n^{-1} \in \omega_G$ for some $n \in N$ and $\gamma \in \omega_\Gamma$, then $n^* \in C^*$ (cf. Harish-Chandra, A formula for semisimple Lie groups, Am. J. Math. 79 (1957), 733-760, Cor. of Lemma 4 for the real case). Let $\mathfrak{n}$ be the Lie algebra of N, $\mathfrak{n} = \sum_{\alpha\in\Phi} \mathfrak{n}_\alpha$ where $\Phi = \Sigma(P, \Gamma)$. For $\alpha \in \Phi$ let ξ_α be the corresponding character of Γ. Fix $\gamma_o \in \Gamma$. Put $\Phi_1 = \{\alpha \in \Phi : \xi_\alpha(\gamma_o) = 1\}$ and let Φ_2 be the complement of Φ_1 in Φ. Define $\mathfrak{n}_1 = \sum_{\alpha\in\Phi_1} \mathfrak{n}_\alpha$, $\mathfrak{n}_2 = \sum_{\alpha\in\Phi_2} \mathfrak{n}_\alpha$. The space $\mathfrak{n}_1$ is a subalgebra of $\mathfrak{n}$.

For each root $\alpha \in \Phi$ let N_α be the subgroup of N characterized by the existence of an Ω-isomorphism $\theta_\alpha : \underset{\sim}{G}_a \longrightarrow \underset{\sim}{N}_\alpha$ ($\underset{\sim}{G}_a$ = the additive group of $\bar{\Omega}$) such that $\gamma\theta_\alpha(x)\gamma^{-1} = \theta_\alpha(\xi_\alpha(\gamma)x)$ $(x \in \underset{\sim}{G}_a,\ \gamma \in \underset{\sim}{\Gamma})$. Let $\underset{\sim}{N}_1$ be the connected Ω-subgroup of $\underset{\sim}{N}$ with Lie algebra $\underset{\sim}{\mathfrak{n}}_1$. Put $N_1 = \underset{\sim}{N}_1 \cap G$. Observe that $\underset{\sim}{N}_1 = \underset{\sim}{N} \cap \underset{\sim}{\Xi}$, where $\underset{\sim}{\Xi}$ is the centralizer of γ_o in $\underset{\sim}{G}$. Furthermore $\underset{\sim}{N}_1 = \prod_{\alpha\in\Phi_1} \underset{\sim}{N}_\alpha$ as Ω-varieties, in any order on Φ_1. Let $\alpha_1 < \alpha_2 < \ldots < \alpha_r$ be all the roots in Φ. Put $\mathfrak{n}^{(k)} = \sum_{k\le i\le r} \mathfrak{n}_{\alpha_i}$ $(k \ge 1)$. Then obviously $\mathfrak{n}^{(k)}$ is an ideal in $\mathfrak{n}$, $\mathfrak{n}^{(1)} = \mathfrak{n}$, $[\mathfrak{n}, \mathfrak{n}^{(k)}] \subset \mathfrak{n}^{(k+1)}$ $(k \ge 1)$ and $\mathfrak{n}^{(k)} = \{0\}$ if k is sufficiently large. Furthermore

$$\mathfrak{n}^{(k)} = \mathfrak{n}^{(k)} \cap \mathfrak{n}_1 + \mathfrak{n}^{(k)} \cap \mathfrak{n}_2$$

for all $k \ge 1$. Let $\underset{\sim}{N}^{(k)}$ be the connected normal Ω-subgroup of $\underset{\sim}{N}$ with Lie algebra $\underset{\sim}{\mathfrak{n}}^{(k)}$. Put $N^{(k)} = \underset{\sim}{N}^{(k)} \cap G$. Then $[N, N^{(k)}] \subset N^{(k+1)}$ and

$N^{(k)}/N^{(k+1)} \simeq N_{\alpha_k}$ for all $k \geq 1$. According to the above order on Φ, we put $N_2 = \prod_{\alpha\in\Phi_2} N_\alpha$. Then $N = N_1 \cdot N_2$ and also $N^{(k)} = (N^{(k)} \cap N_1)\cdot(N^{(k)} \cap N_2)$ for all $k \geq 1$. Choose a compact neighborhood ω_Γ of γ_o in Γ such that, if $\alpha \in \Phi_2$, then ξ_α never takes the value 1 on ω_Γ. Now write $n = n_2 n_1$ $(n \in N,\ n_1 \in N_1,\ n_2 \in N_2)$. Then $n\gamma n^{-1} = n_2 n_1 \gamma n_1^{-1} n_2^{-1} = n_2 \gamma n_1' n_2^{-1}$ where $n_1' = n_1\gamma n_1^{-1}\gamma^{-1} \in N_1$. Since $N_1 \subset \Xi$, we have to show: if $n_2\gamma n_1 n_2^{-1} \in \omega_G$ for some $n_1 \in N_1$, $n_2 \in N_2$, $\gamma \in \omega_\Gamma$, then n_2 remains bounded. To show this, we apply induction: assuming that the assertion is true for $n \in N^{(k+1)}$ $(n = n_2 n_1)$, we shall prove it for $N^{(k)}$. At any rate, the assertion holds for $N^{(r)} = N_{\alpha_r}$. Indeed, we may assume that $\alpha_r \in \Phi_2$. So

$n_2\gamma n_1 n_2^{-1} = n_2 \gamma n_2^{-1} = \gamma(\gamma^{-1} n_2 \gamma n_2^{-1}) = \gamma(\theta_{\alpha_r}((\xi_\alpha(\gamma) - 1)x)$ where $\theta_{\alpha_r}(x) = n_2$.

The choice of ω_Γ yields the desired result. Now assume $n \in N^{(k)}$. Write $n = n_2 n_1$ with $n_2 \in N^{(k)} \cap N_2$, $n_1 \in N^{(k)} \cap N_1$. Furthermore

$n_1 = \nu_1\mu_1$ $(\mu_1 \in N^{(k+1)} \cap N_1,\ \nu_1 \in N_{\alpha_k} \cap N_1)$, $n_2 = \nu_2\mu_2$

$(\mu_2 \in N^{(k+1)} \cap N_2,\ \nu_2 \in N_{\alpha_k} \cap N_2)$. We may assume that $\alpha_k \in \Phi_2$, so $\nu_1 = 1$.

Then $n_2\gamma n_1 n_2^{-1} = \nu_2(\mu_2\gamma\mu_1\mu_2^{-1})\nu_2^{-1} = \gamma(\gamma^{-1}\nu_2\gamma\nu_2^{-1})[\nu_2(\gamma^{-1}\mu_2\gamma\mu_1\mu_2^{-1})\nu_2^{-1}]$.

Now $\nu_2(\gamma^{-1}\mu_2\gamma\mu_1\mu_2^{-1})\nu_2^{-1} \in N^{(k+1)}$ and $\gamma^{-1}\nu_2\gamma\nu_2^{-1} \in N_{\alpha_k}$. Hence both terms are bounded. This implies that ν_2 is bounded (as before in the case $N^{(r)} = N_{\alpha_r}$) and hence, by induction hypothesis, μ_2 is bounded. Therefore n_2 is bounded. This proves our assertion and makes the proof of Lemma 19 complete.

For $x \in G$, write $\det(t - 1 + \mathrm{Ad}(x)) = D(x)t^{\ell} + \ldots$ (terms of higher degree), where t is an indeterminate and $\ell = \operatorname{rank} \underset{\sim}{G}$. We call x <u>regular</u> if $D(x) \neq 0$. Denote by G' the set of regular elements of G. Then G' is an open and dense subset of G, whose complement is of Haar measure zero.

Let Γ again be a Cartan subgroup of G. Put $\Gamma' = \Gamma \cap G'$. It follows easily that Γ' is open and dense in Γ. Let A be the maximal split torus in Γ.

Corollary. <u>Given a compact set</u> ω_Γ <u>in</u> Γ' <u>and a compact set</u> $\omega_G \subset G$, <u>we can choose a compact set</u> $\overline{C}$ <u>in</u> G/A <u>such that</u>

$$(\omega_\Gamma)^x \cap \omega_G Z = \phi$$

<u>unless</u> $\overline{x} \in \overline{C}$.

<u>Proof.</u> The corollary follows from Lemma 19 modulo the following observations. Let Ξ be the centralizer of $\gamma_o \in \Gamma'$ in G. Then the index of Γ in Ξ is finite. Moreover $\Gamma = AB$, where B is a compact torus. This relates G/Ξ to G/A.

Put $G_\Gamma = (\Gamma')^G$, where Γ is as above.

Lemma 20. <u>The mapping</u> $\psi : (x, \gamma) \longmapsto \gamma^x$ <u>of</u> $G \times \Gamma'$ <u>onto</u> G_Γ <u>is submersive.</u>

The proof is the same as in the real case (cf. Harish-Chandra, Invariant eigendistributions on a semisimple Lie group, Trans. A.M.S. 119 (1965), 457-508, Lemma 14).

It follows that G_Γ is open in G. Let dx, $d\gamma$ denote the Haar measures on G, Γ respectively. Observe that $dx d\gamma$ and dx are volume elements on $G \times \Gamma'$ and G_Γ respectively in the sense of §2. So Theorem 11 is applicable with $M = G \times \Gamma'$, $N = G_\Gamma$. For every $\alpha \in C_c^\infty(G \times \Gamma')$ there exists exactly one function $f_\alpha \in C_c^\infty(G_\Gamma)$ such that

$$\int_{G \times \Gamma} \alpha(x, \gamma) F(\gamma^x) dx d\gamma = \int_G f_\alpha(x) F(x) dx$$

for all $F \in C_c^\infty(G_\Gamma)$. For any distribution T on G_Γ let τ be the distribution on $G \times \Gamma'$ given by $\tau(\alpha) = T(f_\alpha)$ $(\alpha \in C_c^\infty(G \times \Gamma'))$. Since $\alpha \longmapsto f_\alpha$ is surjective, τ is uniquely determined by T. Notice that G operates on G_Γ by $y^g = gyg^{-1}$ $(y, g \in G)$.

Lemma 21. Let T be an invariant distribution on $G_{\Gamma'}$. Then there exists a unique distribution σ_T on Γ' such that $T(f_a) = \sigma_T(\beta_a)$ $(a \in C_c^{\infty}(G \times \Gamma'))$, where

$$\beta_a(\gamma) = \int_G a(x, \gamma)dx \qquad (\gamma \in \Gamma') .$$

Proof. For $a \in C_c^{\infty}(G \times \Gamma')$, define $\lambda(y)a(x, \gamma) = a(y^{-1}x, \gamma)$ $(x, y \in G, \gamma \in \Gamma')$. Fix $y \in G$ and put $a' = \lambda(y)a$. Then we have

$$\int f_{a'}(x)F(x)dx = \int a'(x, \gamma)F(\gamma^x)dxd\gamma$$
$$= \int a(y^{-1}x, \gamma)F(\gamma^x)dxd\gamma = \int a(x, \gamma)F(\gamma^{yx})dxd\gamma = \int a(x, y)F^{y^{-1}}(\gamma^x)dxd\gamma$$
$$= \int f_a(x)F^{y^{-1}}(x)dx = \int (f_a)^y(x)F(x)dx \quad \text{for all } F \in C_c^{\infty}(G_{\Gamma'}) .$$

Hence $f_{\lambda(y)a} = (f_a)^y$. So we get $\tau(\lambda(y)a) = T(f_{\lambda(y)a}) = T(f_a) = \tau(a)$. This means that τ is invariant under left-translations by elements of G. Fix $\beta \in C_c^{\infty}(\Gamma')$ and put $\tau_\beta(\delta) = \tau(\delta \otimes \beta)$ $(\delta \in C_c^{\infty}(G))$. Then $\tau_\beta(\lambda(y)\delta) = \tau_\beta(\delta)$ for all $y \in G$, hence by Lemma 17,

$$\tau_\beta(\delta) = \sigma_T(\beta) \int_G \delta(x)dx$$

where $\sigma_T(\beta)$ is a constant. Now select $\delta_o \in C_c^{\infty}(G)$ such that $\int \delta_o(x)dx = 1$. Then $\sigma_T(\beta) = \tau_\beta(\delta_o) = \tau(\delta_o \otimes \beta)$ $(\beta \in C_c^{\infty}(\Gamma'))$. This shows that the mapping $\beta \longmapsto \sigma_T(\beta)$ is a distribution on Γ'. We have

$$\tau(\delta \otimes \beta) = \sigma_T(\beta) \int \delta(x)dx \qquad (\delta \in C_c^{\infty}(G), \beta \in C_c^{\infty}(\Gamma')) .$$

Since $C_c^{\infty}(G \times \Gamma') = C_c^{\infty}(G) \otimes C_c^{\infty}(\Gamma')$, we get easily

$$\tau(a) = \sigma_T(\beta_a) \qquad (a \in C_c^{\infty}(G \times \Gamma'))$$

where β_a is given by $\beta_a(\gamma) = \int a(x, \gamma)dx$ $(\gamma \in \Gamma')$. Since $\beta_a = \beta$ for $a = \delta_o \otimes \beta$ $(\beta \in C_c^{\infty}(\Gamma'))$, the mapping $a \longmapsto \beta_a$ of $C_c^{\infty}(G \times \Gamma')$ into

$C_c^\infty(\Gamma')$ is surjective. So σ_T is uniquely determined by τ and therefore by T. This completes the proof.

One may compare this proof with that of: Harish-Chandra, T.A.M.S. 119 (1965), 457-508, Lemma 15.

§4. Applications to the characters of the supercuspidal representations.

Let π be a supercuspidal representation on the Hilbert space $\mathcal{H}_\pi$. Choose a K-finite unit vector $\phi \in \mathcal{H}_\pi$ and define

$$\theta(y) = (\phi, \pi(y)\phi) \qquad (y \in G) .$$

Then θ is a locally constant supercusp form. Let Θ_π be the invariant distribution on G given by

$$\Theta_\pi(f) = \operatorname{tr} \pi(f) \qquad (f \in C_c^\infty(G)) .$$

Let us compute σ_T for $T = \Theta_\pi$ on G_Γ. With the notations of Lemma 21 we have (cf. Theorem 9),

$$\sigma_T(\beta_a) = d(\pi) \int_{G/Z} dx^* \int_{G_\Gamma} f_a(y)\theta(y^x)dy .$$

By the corollary of Theorem 11, we get

$$\sigma_T(\beta_a) = d(\pi) \int_{G/Z} dx^* \int_{G\times\Gamma'} a(y, \gamma)\theta(\gamma^{xy})dyd\gamma \qquad (a \in C_c^\infty(G\times\Gamma')) .$$

Now choose an open compact subgroup K_o of G and put

$$\delta_o = \frac{\text{characteristic function of } K_o}{\text{measure of } K_o} .$$

Let $a = \delta_o \otimes \beta$ $(\beta \in C_c^\infty(\Gamma'))$. Then $\beta_a = \beta$ and

$$\sigma_T(\beta) = d(\pi) \int_{G/Z} dx^* \int_{\Gamma'\times K_o} \beta(\gamma)\theta(\gamma^{xk})dk \qquad (\beta \in C_c^\infty(\Gamma')) ,$$

where dk is a normalized Haar measure on K_o: $\int_{K_o} dk = 1$.

Lemma 22. Let P be a parabolic subgroup with unipotent radical N. Denote by $\mathfrak{n}$ the Lie algebra of N.

(i) For any semisimple element $\gamma \in P$ such that

$$\Delta(\gamma) = \det(\mathrm{Ad}(\gamma^{-1}) - 1)_{\mathfrak{n}} \neq 0 ,$$

the mapping $n \longmapsto \gamma n \gamma^{-1} n^{-1}$ is an Ω-isomorphism of algebraic varieties of N onto N.

(ii) Denote by dn a Haar measure on N. Then for all $f \in C_c(G)$ and all semisimple $\gamma \in P$ with $\Delta(\gamma) \neq 0$,

$$\int_N f(\gamma n) dn = |\det(\mathrm{Ad}(\gamma^{-1}) - 1)_{\mathfrak{n}}| \int_N f(\gamma^n) dn .$$

Proof. Since the centralizer Ξ of γ in N is connected and $\Delta(\gamma) \neq 0$, it follows that $\Xi = \{1\}$. The first assertion now follows from Borel-Tits, 11.1 (p. 131). The second assertion is easy in char $\Omega = 0$ and can be proved for instance as in: Harish-Chandra, Two theorems on semisimple Lie groups, Ann. Math. 83 (1966), 74-128, Lemma 11 . To cover the case of positive characteristic, we give a proof. Choose a filtration $N = N_1 \supset N_2 \supset \ldots \supset N_r \supset N_{r+1} = \{e\}$ of N such that $[N_i, N_i] \subset N_{i+1}$ and such that $\mathrm{Ad}(\gamma)N_i \subset N_i$ $(1 \leq i \leq r)$. Denote by $\mathfrak{n}_i$ the Lie algebra of N_i $(1 \leq i \leq r)$. It is not difficult to show that

$$\int_N f(n) dn = \int_{N_1/N_2} \ldots \int_{N_r/N_{r+1}} f(\dot{n}_1 \ldots \dot{n}_r) d\dot{n}_1 \ldots d\dot{n}_r \qquad (f \in C_c(N)) .$$

Now applying that N_i/N_{i+1} $(1 \leq i \leq r)$ is abelian and $\dot{n}_i \longmapsto \gamma^{-1} \dot{n}_i \gamma \dot{n}_i^{-1}$ $(\dot{n}_i \in N_i/N_{i+1})$ is an homomorphism, hence an isomorphism by (i), of the group N_i/N_{i+1}, and keeping in mind that we are dealing with unipotent groups, we obtain easily the desired result:

$$\int_N f(\gamma^{-1} n \gamma n^{-1}) = \prod_{1 \le i \le r} |\det(\mathrm{Ad}(\gamma^{-1}) - 1)_{\mathfrak{n}_i/\mathfrak{n}_{i+1}}| \int_N f(n)dn$$

$$= |\det(\mathrm{Ad}(\gamma^{-1}) - 1)_{\mathfrak{n}}| \int_N f(n)dn \qquad (f \in C_c(N)) .$$

Lemma 23. Let ω_Γ be a compact subset of Γ'. There exists a compact set $\omega \subset G$ such that $\int_{K_o} \theta(\gamma^{xk})dk = 0$ for all $\gamma \in \omega_\Gamma$ unless $x \in \omega Z$.

Proof. Since θ is a supercusp form, there exists a compact set ω_G in G such that $\mathrm{Supp}\ \theta \subset \omega_G Z$. By the corollary of Lemma 19, there exists a compact set $\overline{C}$ in G/A such that

$$(\omega_\Gamma)^x \cap \omega_G Z = \emptyset$$

for $\bar{x}$ outside $\overline{C}$. Hence $(\omega_\Gamma)^x \cap \mathrm{Supp}\ \theta = \emptyset$ unless $\bar{x} \in \overline{C}$. Let C denote the complete inverse image of $\overline{C}$ in G. Put $g_\gamma(x) = \theta(\gamma^x)$ $(x \in G,\ \gamma \in \omega_\Gamma)$. Then for all $\gamma \in \omega_\Gamma$

$$\mathrm{Supp}\ g_\gamma \subset C \text{ and } g_\gamma(xa) = g_\gamma(x) \qquad (x \in G,\ a \in A) .$$

Let $\mathcal{P}'$ be the set of cuspidal pairs (P', A') with $A' \subset A$, $P' \neq G$. For any pair $(P', A') \in \mathcal{P}'$, $P' = M'N'$, we have $\omega_\Gamma \subset M'$ and

$$\int_{N'} g_\gamma(xn)dn = \int_{N'} \theta(\gamma^{xn})dn = \int_{N'} \theta^{x^{-1}}(\gamma^n)dn$$

$$= |\det(\mathrm{Ad}(\gamma^{-1}) - 1)_{\mathfrak{n}'}|^{-1} \int_{N'} \theta^{x^{-1}}(\gamma n)dn \qquad (\gamma \in \omega_\Gamma) .$$

Here $\mathfrak{n}'$ is the Lie algebra of N'. Since γ is regular, $\det(\mathrm{Ad}(\gamma^{-1}) - 1)_{\mathfrak{n}'} \neq 0$. Furthermore $\theta^{x^{-1}}$ is a supercusp form for all $x \in G$. Hence

$$\int_{N'} g_\gamma(xn)dn = 0 \qquad (x \in G) .$$

All conditions of Theorem 10 are now fulfilled. So we can choose a compact set $\omega \subset G$ with the following properties. There is an open subgroup $K_1 \subset K_o$

such that for all $\gamma \in \omega_\Gamma$,

$$\int_{K_1} g_\gamma(xk)dk = 0 \quad \text{unless} \quad x \in \omega Z \qquad (x \in G) .$$

But then also

$$\int_{K_o} g_\gamma(xk)dk = \sum_{1 \leq i \leq r} \int_{K_1} g_\gamma(xk_i k)dk = 0 \qquad (x \in G)$$

where $(k_i)_{1 \leq i \leq r}$ is a complete set of representatives of K_o mod K_1. Hence $\int_{K_o} \theta(\gamma^{xk})dk = 0$ for all $\gamma \in \omega_\Gamma$, unless $x \in \omega Z$. This completes the proof.

Define $F_\Gamma(\gamma) = d(\pi) \int_{G/Z} dx^* \int_{K_o} \theta(\gamma^{xk})dk \quad (\gamma \in \Gamma')$. By Lemma 23 this integral exists and F_Γ is locally constant. We have

$$\sigma_T(\beta) = \int_\Gamma \beta(\gamma)F_\Gamma(\gamma)d\gamma \qquad (\beta \in C_c^\infty(\Gamma')) .$$

Observe that the definition of F_Γ does not depend on the choice of K_o. For x_o in the normalizer of Γ, we get

$$F_\Gamma(\gamma^{x_o}) = d(\pi) \int_{G/Z} dx^* \int_{K_o} \theta(\gamma^{xkx_o})dk$$

$$= d(\pi) \int_{G/Z} dx^* \int_{K_o^{x_o}} \theta(\gamma^{xk})dk = F_\Gamma(\gamma) \qquad (\gamma \in \Gamma')$$

where $K_o^{x_o} = x_o K_o x_o^{-1}$.

Theorem 12. Let π be a supercuspidal representation and Θ_π its character. Then there exists a locally constant function F on G' such that $\Theta_\pi = F$ on G'.

Proof. It is obviously sufficient to prove the theorem locally. Fix $\gamma_o \in G'$ and let Γ be the Cartan subgroup which contains γ_o. Put $\overline{G} = G/\Gamma$.

The mapping $\bar{\psi} : (\bar{x}, \gamma) \longmapsto \gamma^{\bar{x}} = x\gamma x^{-1}$ of $\bar{G} \times \Gamma'$ onto G_Γ is locally an analytic homeomorphism. Choose open compact neighborhoods ω_Γ and $\bar{G}_o$ of γ_o and $\bar{1}$ in Γ' and $\bar{G}$ respectively such that $\bar{\psi}$ is an analytic homeomorphism of $\bar{G}_o \times \omega_\Gamma$ onto $\bar{\psi}(\bar{G}_o \times \omega_\Gamma) = V$. V is an open compact neighborhood of γ_o in G_Γ. Define

$$F(\gamma^{\bar{x}}) = F_\Gamma(\gamma) \qquad (\gamma \in \omega_\Gamma, \ \bar{x} \in \bar{G}_o)$$

where F_Γ is defined above. This definition is meaningful. Let G_o be the inverse image of $\bar{G}_o$ in G. For any $\alpha \in C_c^\infty(G \times \Gamma')$ with $\operatorname{Supp} \alpha \subset G_o \times \omega_\Gamma$ we have $\operatorname{Supp} f_\alpha \subset (\omega_\Gamma)^{G_o} = V$ and

$$\int f_\alpha(x)F(x)dx = \int \alpha(x, \gamma)F(\gamma^x)dxd\gamma$$
$$= \int \beta_\alpha(\gamma)F_\Gamma(\gamma)d_\gamma = \sigma_T(\beta_\alpha) = \Theta(f_\alpha) \quad \text{(Lemma 21)} .$$

Hence $\Theta = F$ on V. It is easy to see that F is locally constant since F_Γ is. This completes the proof.

Notice that F is given by $F(\gamma) = F_\Gamma(\gamma)$ for $\gamma \in G' \cap \Gamma$, where Γ stands for Cartan subgroups of G. Furthermore $F^x = F$ for all $x \in G$ since Θ is an invariant distribution. So it is sufficient to consider a complete set of non-conjugate Cartan subgroups to compute F. Observe that this set may be infinite. Take for instance: char $\Omega = 2$, $G = SL(2, \Omega)$. In this case any quadratic extension of Ω provides one or more (compact) Cartan subgroups, while $[\Omega^* : (\Omega^*)^2] = \infty$. If char $\Omega = 0$, this set is finite for all reductive $\mathfrak{p}$-adic groups G, defined over Ω.

The next lemma is suggested by the results of Sally and Shalika in case $SL(2, \Omega)$. It shows that the characters of the supercuspidal representations not only vanish fast at ∞, but are zero outside a compact set on Γ (mod Z).

Lemma 24. Let Γ be a Cartan subgroup of G. There exists a compact set ω_Γ in Γ such that $F_\Gamma(\gamma) = 0$ unless $\gamma \in \omega_\Gamma Z \cap \Gamma'$.

Proof. It follows from the definition of F_Γ that $F_\Gamma(\gamma) = 0$ if γ is outside the closure of the set $(\mathrm{Supp}\ \theta)^G \cap \Gamma$. We shall show that $\omega^G \cap \Gamma$ is relatively compact mod Z in G for all sets $\omega \subset G$ which are compact mod Z. Let A be the maximal split torus in Γ. $\Gamma = AB$ where B is a compact torus. Let P be a parabolic subgroup of G with A as a split component. We have $G = K P = KNM$ if $P = MN = NM$ is the Levi-decomposition of (P, A). Take $\gamma \in \omega^G \cap \Gamma$. Then $\gamma^x \in \omega$ for some $x \in G$. Write $x = knm$. Then $\gamma^{nm} \in \omega^{k^{-1}} \subset \omega^K = \omega'$. Since $\Gamma \subset M$ we see, writing $\gamma^{nm} = \gamma^m(\gamma^{-m} n \gamma^m n^{-1})$, that γ^m stays in a compact set mod Z. Put $\gamma = ab$ $(a \in A,\ b \in B)$. So ab^m remains bounded. Hence for any root α of G w.r.t. A, $|\det(\mathrm{Ad}(ab^m))_{\mathfrak{g}_\alpha}|$ is bounded, where $\mathfrak{g}_\alpha \subset \mathfrak{g}$ is the root space belonging to α and $\mathfrak{g}$ is the Lie algebra of G. But $|\det(\mathrm{Ad}(ab^m))_{\mathfrak{g}_\alpha}| = |\det(\mathrm{Ad}(ab))_{\mathfrak{g}_\alpha}| = |\xi_\alpha(a)|^{m_\alpha}$, where ξ_α is the character corresponding to α and $m_\alpha = \dim \mathfrak{g}_\alpha$. Hence a remains in a compact set mod Z and therefore γ too. This proves the lemma.

Part VI. The mapping "F_f" (char $\Omega = 0$).

Let π be a supercuspidal representation and let Θ_π denote its character. In Part V we have shown that the restriction of Θ_π to G' is a locally constant function, which we denoted by F. Put $F = 0$ outside G'. We are going to prove that F is locally summable on G and $\Theta = F$. The proof requires a long preparation. This Part contains one of the main steps in the proof.

We often shall replace the group G by its Lie algebra $\mathfrak{g}$. Apart from reasons, which will be clear by looking at the results used in the proofs, the absence of a "good" mapping "exp" on $\mathfrak{g}$ in case of positive characteristic, forces us to assume from now on that char $\Omega = 0$.

§1. Introduction and elementary properties of the mapping "Φ_f".

As before, let $\underset{\sim}{G}$ be a connected reductive linear algebraic group, defined over Ω, and let G be the group of Ω-rational points of $\underset{\sim}{G}$. Denote by $\mathcal{F}_\Omega(\underset{\sim}{G})$ the field of rational functions defined over Ω on $\underset{\sim}{G}$. The Lie algebra $\mathfrak{g}$ of all left-invariant Ω-derivations of $\mathcal{F}_\Omega(\underset{\sim}{G})$ is called the Lie algebra of G. We have $\dim_\Omega \mathfrak{g} = \dim \underset{\sim}{G}$. Furthermore $\mathfrak{g}$ is reductive. For $X \in \mathfrak{g}$, put

$$\det(t - \operatorname{ad} X) = \eta(X)t^{\ell} + \ldots \qquad \text{(terms of higher degree in } t)$$

where t is an indeterminate. Here $\ell = \operatorname{rank} \mathfrak{g} = \operatorname{rank} \underset{\sim}{G}$. X is called <u>regular</u> if $\eta(X) \neq 0$. The set of regular elements of $\mathfrak{g}$ is denoted by $\mathfrak{g}'$.

In this Part, let A stand for a Cartan subgroup of G. Let A' be the set of regular elements of A. For any $h \in C_c^\infty(G)$, put

$$F_h(a) = |D(a)|^{1/2} \int_{G/A} h(a^{x^*})dx^* .$$

This integral exists for $a \in A'$, even in case Supp h is compact mod Z (Lemma 19). Our main aim in this Part is to prove that F_h is locally <u>bounded</u> on A' (cf. §8 for a precise statement). We pass to the Lie algebra

$\mathfrak{g}$ of G. Let $\mathfrak{a}$ be the Cartan subalgebra of $\mathfrak{g}$ corresponding to A. Put $\mathfrak{a}' = \mathfrak{a} \cap \mathfrak{g}'$. If $\mathcal{X} \subset \mathfrak{g}$ and $\mathcal{G} \subset G$, we write $\mathcal{G}\mathcal{X}$ or $\mathcal{X}^{\mathcal{G}}$ in place of $\mathrm{Ad}(\mathcal{G})\mathcal{X}$. Similarly if $X \in \mathfrak{g}$, $x \in G$ we write xX or X^x instead of $\mathrm{Ad}(x)X$. Let f be any function in $C_c^{\infty}(\mathfrak{g})$. Define

$$\Phi_f(H) = |\eta(H)|^{1/2} \int_{G/A} f(x^* H) dx^* \qquad (H \in \mathfrak{a}') .$$

We shall soon show that this integral exists. We shall prove:

Theorem 13. <u>Fix</u> $f \in C_c^{\infty}(\mathfrak{g})$. <u>Then</u> $\sup_{H \in \mathfrak{a}'} |\Phi_f(H)| < +\infty$.

The connection between the boundedness of Φ_f and F_h will be treated later on.

The existence of the integral, defining Φ_f, is guaranteed by the following lemma.

Lemma 25. <u>Fix</u> $H_o \in \mathfrak{a}$. <u>Let</u> Ξ <u>be the centralizer of</u> H_o <u>in</u> G. <u>Then there exists a neighborhood</u> V <u>of</u> H_o <u>in</u> $\mathfrak{a}$ <u>with the following property. Given any compact set</u> $\omega \subset \mathfrak{g}$, <u>we can choose a compact set</u> $\overline{C}$ <u>in</u> $\overline{G} = G/\Xi$ <u>such that</u> $V^x \cap \omega = \emptyset$ <u>unless</u> $\bar{x} \in \overline{C}$.

The proof of this lemma is exactly the same as in the real case (cf. Harish-Chandra, Some results on an invariant integral on a semisimple Lie algebra, Ann. Math. 80 (1964), 551-593, Lemma 7). Only for the convenience of the reader we shall present it here.

<u>Proof of Lemma</u> 25. For the purpose of this lemma we may assume that $\mathfrak{g}$ is semisimple. Choose an open neighborhood U of 0 in $\mathfrak{g}$ such that

(i) $U = U^x$ for all $x \in G$.

(ii) "exp" is defined on U (we use the formula $\exp H^x = (\exp H)^x$ for defining "exp" on U).

(iii) "exp" is injective and submersive on U.

(Cf. Harish-Chandra, Am. J. Math. 79 (1957), 733-760, Lemma 11, where the real case is treated.)

Without loss of generality we may assume that H_o and the given compact set $\omega \subset \mathfrak{g}$ are both contained in U. Put $a_o = \exp H_o$. Then obviously Ξ = centralizer of a_o in G. Since "exp" is injective on U, we may consider $\exp(V^x \cap \omega)$ in place of $V^x \cap \omega$. But

$$\exp(V^x \cap \omega) = (\exp V)^x \cap \exp \omega$$

and, "exp" being submersive on U, we are reduced to Lemma 19 for the point $a_o \in G$ and the compact set $\exp \omega \subset G$. This proves Lemma 25.

Notice that $[\Xi : A] < +\infty$ if $H_o \in \mathfrak{a}'$.

It will be sufficient to prove the theorem in case G (and $\mathfrak{g}$) is semi-simple. To show this, let G_o be an open subgroup of G of finite index and let A be a Cartan subgroup of G. Put $A_o = A \cap G_o$. We have a canonical injection $G_o/A_o \longrightarrow G/A$ and it is easy to see that this map is open and continuous. We identify G_o/A_o in the obvious way with an open subspace of G/A. For any $f \in C_c^\infty(\mathfrak{g})$, put

$$\Phi_f^o(H) = |\eta(H)|^{1/2} \int_{G_o/A_o} f(x^* H)dx^* \qquad (H \in \mathfrak{a}')$$

where $\mathfrak{a}$ is the Lie algebra of A.

Lemma 26. <u>Assume Φ_f^o is bounded on $\mathfrak{a}'$ for all $f \in C_c^\infty(\mathfrak{g})$ and every choice of</u> $(A, \mathfrak{a})$. <u>Then the same is true for Φ_f, and conversely.</u>

<u>Proof.</u> The "converse" is obvious, since $|\Phi_f^o| \leq \Phi_{|f|}$. Normalize the Haar measures dx and da on G and A respectively, and the invariant measure dx^* on G/A in such a way that $dx = dx^* da$. For any $y \in G$ we define a Haar measure da^y on A^y by the formula

$$\int f(a^y)da^y = \int f(a^{y^{-1}})da \qquad (f \in C_c(A^y)) .$$

Normalize the invariant measure $d_y x^*$ on G/A^y such that $dx = d_y x^* \, da^y$. Then we have for any function $g \in C_c(G/A)$:

$$\int_{G/A} g(x^*)dx^* = \sum_{y \in G_o \backslash G/A} \int_{G_o yA/A} g(x^*)dx^* = \sum_{y \in G_o \backslash G/A} \int_{G_o/G_o \cap A^y} g(x^* y)d_y x^*$$

Now fix $f \in C_c^\infty(\mathfrak{g})$. We get

$$|\eta(H)|^{1/2} \int_{G/A} f(x^* H)dx^* = \sum_{y \in G_o \backslash G/A} |\eta(yH)|^{1/2} \int_{G_o/G_o \cap A^y} f(x^*(yH))d_y x^*$$

and this leads straightforward to the desired result.

Write $\mathfrak{g} = \mathfrak{z} + \mathfrak{g}_1$, where $\mathfrak{z}$ is the center of $\mathfrak{g}$ and $\mathfrak{g}_1$ the derived Lie algebra, which is semisimple. On the other hand $\underset{\sim}{G} = \underset{\sim}{T}.\underset{\sim}{G}_1$, where $\underset{\sim}{T}$ is a central Ω-torus and $\underset{\sim}{G}_1$ is the derived group of $\underset{\sim}{G}$, which is defined over Ω and semisimple. Furthermore $\underset{\sim}{T} \cap \underset{\sim}{G}_1$ is finite. Put $T = \underset{\sim}{T} \cap G$, $G_1 = \underset{\sim}{G}_1 \cap G$ and $G_o = T.G_1$. Then G_o is an open subgroup of G (in the $\mathfrak{p}$-adic topology) of finite index (char $\Omega = 0$!). Moreover $\mathfrak{g}_1$ is the Lie algebra of G_1.

Let A be a Cartan subgroup of G. Then $A_1 = A \cap G_1$ is a Cartan subgroup of G_1. If $\mathfrak{a}$ is the Lie algebra of A, then $\mathfrak{a}_1 = \mathfrak{a} \cap \mathfrak{g}_1$ is the Lie algebra of A_1. Furthermore $\mathfrak{a} = \mathfrak{z} + \mathfrak{a}_1$.

Lemma 27. <u>Assume that Theorem 13 is true for</u> G_1. <u>Then it is true for</u> G.

<u>Proof.</u> Let A be a Cartan subgroup of G with Lie algebra $\mathfrak{a}$. Fix $f \in C_c^\infty(\mathfrak{g})$. By Lemma 26 it suffices to show that

$$\Phi_f^o(H) = |\eta(H)|^{1/2} \int_{G_o/A \cap G_o} f(x^* H)dx^* \qquad (H \in \mathfrak{a}')$$

is bounded on $\mathfrak{a}'$. Since $C_c^\infty(\mathfrak{g}) = C_c^\infty(\mathfrak{g}_1) \otimes C_c^\infty(\mathfrak{z})$ we may assume that f is of the form $f = f_1 \otimes f_2$ $(f_1 \in C_c^\infty(\mathfrak{g}_1),\ f_2 \in C_c^\infty(\mathfrak{z}))$. Furthermore, since

$T \subset A \cap G_o$ we have $G_o/A \cap G_o = G_1/A_1$, where $A_1 = A \cap G_1$. Writing $H = H_1 + Z$ ($H \in \mathfrak{a}$, $H_1 \in \mathfrak{a}_1 = \mathfrak{a} \cap \mathfrak{g}_1$, $Z \in \mathfrak{z}$) and observing that $\eta(H_1 + Z) = \eta(H_1) = \eta_1(H_1)$, where η_1 is the function "η" belonging to $\mathfrak{g}_1$, we obtain

$$|\Phi_f^{\,o}(H)| \leq |\eta_1(H_1)|^{1/2} \int_{G_1/A_1} |f_1|(x^* H_1)|f_2|(x^* Z)dx^*$$

$$\leq \sup_{Z \in \mathfrak{z}} |f_2(Z)| \cdot |\eta_1(H_1)|^{1/2} \int_{G_1/A_1} |f_1|(x^* H_1)dx^* .$$

So $\sup_{H \in \mathfrak{a}'} |\Phi_f^{\,o}(H)| < +\infty$ by the assumption on G_1. This proves the lemma.

By Lemma 27, the proof of Theorem 13 reduces to the case where G and $\mathfrak{g}$ are semisimple.

We conclude this paragraph with a lemma and corollary concerning the support of Φ_f.

Lemma 28. <u>Fix a compact subset $\omega \subset \mathfrak{g}$. The set of all $H \in \mathfrak{a}$ satisfying $H \in Cl(\omega^G)$ is relatively compact in $\mathfrak{g}$.</u>

<u>Proof.</u> Write $\omega \subset \omega_1 + \omega_2$, where $\omega_1 \subset \mathfrak{g}_1$, $\omega_2 \subset \mathfrak{z}$ and ω_1, ω_2 compact. Observing that $\mathfrak{a} = \mathfrak{a}_1 + \mathfrak{z}$, where $\mathfrak{a}_1 = \mathfrak{a} \cap \mathfrak{g}_1$ and $\omega_1^{\,G} \subset \mathfrak{g}_1$, $\omega_2^{\,G} = \omega_2$, we may restrict ourselves to the case of semisimple $\mathfrak{g}$. Denote by J the algebra of all G-invariant polynomials on $\mathfrak{g}$. For $p \in J$ we have $|p(H)| \leq \sup_{X \in \omega} |p(X)|$ if $H \in Cl(\omega^G)$. Denote by S the set of all $H \in \mathfrak{a}$ such that $H \in Cl(\omega^G)$. Each $p \in J$ is bounded on S. Now consider for $X \in \mathfrak{g}$,

(*) $\quad \det(t - \operatorname{ad} X) = t^n + p_1(X)t^{n-1} + \ldots + p_{n-\ell}(X)t^{\ell} \qquad (p_{n-\ell} = \eta,\ \ell = \operatorname{rank} \mathfrak{g})$

We have $p_1, \ldots, p_{n-\ell} \in J$, hence they are bounded on S. The Lie algebra $\operatorname{ad}(\mathfrak{a})$ splits over a finite field extension L of Ω. L is naturally endowed with the $\mathfrak{p}$-adic valuation, which extends that of Ω. It follows easily from (*) that the roots of $(\mathfrak{g} \otimes L, \mathfrak{a} \otimes L)$ are bounded on S. Hence S is bounded in $\mathfrak{g}$. This completes the proof.

Corollary. Fix $f \in C_c^\infty(\mathfrak{g})$. Then $\Phi_f(H) = 0$ for $H \in \mathfrak{a}'$ lying outside a compact subset of $\mathfrak{a}$.

Proof. Suppose $H_o \in \mathfrak{a}'$ satisfies $\Phi_f(H) \neq 0$. Then every neighborhood of H_o meets $(\mathrm{Supp}\, f)^G$. Hence $H_o \in \mathfrak{a} \cap \mathrm{Cl}((\mathrm{Supp}\, f)^G)$. But by Lemma 28, the set of all such points H_o is relatively compact. This proves the corollary.

§2. The first step in the proof of Theorem 13.

Theorem 13 is proved by induction on $\dim G = \dim \underset{\sim}{G}$. So suppose the theorem to be true for all (reductive) groups with dimension $m < \dim G$. We shall prove the theorem for G with $\dim G > 0$ which, as we saw, may be assumed to be semisimple. This we will assume from now on. Let A be a Cartan subgroup of G with Lie algebra $\mathfrak{a} \subset \mathfrak{g}$. Put as before,

$$\Phi_f(H) = |\eta(H)|^{1/2} \int_{G/A} f(x^* H)dx^* \qquad (H \in \mathfrak{a}', f \in C_c^\infty(\mathfrak{g})) .$$

Let us denote by $\mathcal{N}$ the set of nilpotent elements of $\mathfrak{g}$. Using the notation of (*) (see Lemma 28), one easily sees that $\mathcal{N}$ is the set of common zeros of the polynomials $p_1, \ldots, p_{n-\ell}$ on $\mathfrak{g}$ ($\ell = \mathrm{rank}\, \mathfrak{g}$).

Lemma 29. Let $f \in C_c^\infty(\mathfrak{g})$ satisfy $\mathrm{Supp}\, f \cap \mathcal{N} = \emptyset$. Then Φ_f is bounded on $\mathfrak{a}'$.

Proof. Fix $f \in C_c^\infty(\mathfrak{g})$. By the corollary of Lemma 28 it is enough to prove: given $H_o \in \mathfrak{a}$, we can choose a neighborhood V of H_o in $\mathfrak{a}$ such that

$$\sup_{H \in V'} |\Phi_f(H)| < +\infty, \quad \text{where } V' = V \cap \mathfrak{g}' .$$

There are two cases to be considered separately.

(i) Assume $H_o \neq 0$. Let $\underset{\sim}{\Xi}$ be the centralizer of H_o in $\underset{\sim}{G}$. $(\underset{\sim}{\Xi})^o$ is a connected, reductive Ω-subgroup of $\underset{\sim}{G}$. Put $\Xi_o = (\underset{\sim}{\Xi})^o \cap G$.

Choose with Lemma 25 a neighborhood V of H_o in $\mathfrak{a}$ such that $V^x \cap \mathrm{Supp}\, f = \emptyset$ unless $\bar{x} \in \bar{C}$, $\bar{C}$ being an open compact subset of G/Ξ_o. Denote by $\mathfrak{z}_{H_o}$ the centralizer of H_o in $\mathfrak{g}$. Then $\mathfrak{z}_{H_o}$ is the Lie algebra of Ξ_o, since H_o is semisimple. Put $\mathfrak{g}_{H_o} = (\mathrm{ad}\, H_o)\mathfrak{g} = [H_o, \mathfrak{g}]$. We have $\mathfrak{g} = \mathfrak{z}_{H_o} + \mathfrak{g}_{H_o}$, $[\mathfrak{z}_{H_o}, \mathfrak{g}_{H_o}] = \mathfrak{g}_{H_o}$, $\mathfrak{a} \subset \mathfrak{z}_{H_o}$. Hence $\mathfrak{a}$ is a Cartan subalgebra of $\mathfrak{z}_{H_o}$. Moreover $A \subset \Xi_o$ and A is a Cartan subgroup of Ξ_o. Since $H_o \neq 0$, we have $\dim \Xi_o < \dim G$. We shall apply the induction hypothesis to Ξ_o.

For any $Z \in \mathfrak{z}_{H_o}$, put $\det(t - \mathrm{ad}\, Z) = \det(t - \mathrm{ad}\, Z)_{\mathfrak{z}_{H_o}} \cdot \det(t - \mathrm{ad}\, Z)_{\mathfrak{g}_{H_o}}$. Define $\nu(Z) = \det(\mathrm{ad}\, Z)_{\mathfrak{g}_{H_o}}$. We get $\eta(Z) = \eta_{\mathfrak{z}_{H_o}}(Z)\nu(Z)$ $(Z \in \mathfrak{z}_{H_o})$. Now $\nu(H_o) \neq 0$.

Let $\mathfrak{z}_o$ be the set of all $Z \in \mathfrak{z}_{H_o}$ satisfying $|\nu(Z)| = |\nu(H_o)|$. $\mathfrak{z}_o$ is an open neighborhood of H_o in $\mathfrak{z}_{H_o}$, invariant under Ξ_o. We may assume $V \subset \mathfrak{z}_o$.

Fix Haar measures dx, da, $d\xi$ on G, A, Ξ_o respectively. Choose invariant measures $d\xi^*$, dx^*, $d\bar{x}$ on Ξ_o/A, G/A and G/Ξ_o respectively such that $d\xi = d\xi^* da$, $dx = dx^* da$, $dx = d\bar{x}d\xi$. We have

$$\int_{G/A} f(x^*H)dx^* = \int_{G/\Xi_o} d\bar{x} \int_{\Xi_o/A} f(x(\xi^*H))d\xi^* \qquad (H \in \mathfrak{a}') .$$

Now take $H \in V' = V \cap \mathfrak{g}'$. Then

$$\bar{x} \longmapsto \int_{\Xi_o/A} f(x(\xi^*H))d\xi^*$$

vanishes outside $\bar{C} \subset G/\Xi_o$, chosen above. Choose $\alpha \in C_c^\infty(G)$ such that

$$\bar{\alpha}(\bar{x}) = \int_{\Xi_o} \alpha(x\xi)d\xi = \begin{cases} 1 & \text{if } \bar{x} \in \bar{C} \\ 0 & \text{elsewhere} \end{cases} \qquad (\bar{x} \in G/\Xi_o) .$$

Put $g(Z) = \int_G a(x)f(xZ)dx \quad (Z \in \mathfrak{z}_{H_o})$. Then clearly, $g \in C_c^\infty(\mathfrak{z}_{H_o})$. Furthermore

$$\int_{\Xi_o/A} g(\xi^* H)d\xi^* = \int_G a(x)dx \int_{\Xi_o/A} f(x(\xi^* H))d\xi^*$$

$$= \int_{G/\Xi_o} \bar{a}(\bar{x})d\bar{x} \int_{\Xi_o/A} f(x(\xi^* H))d\xi^* = \int_{G/A} f(x^* H)dx^* \qquad (H \in V') .$$

We obtain

$$\Phi_f(H) = |\nu(H_o)|^{1/2} |\eta_{\mathfrak{z}_{H_o}}(H)|^{1/2} \int_{\Xi_o/A} g(\xi^* H)d\xi^* \qquad (H \in V') .$$

The induction hypothesis applied to Ξ_o yields: $\sup_{H \in V'} |\Phi_f(H)| < +\infty$.

(ii) Suppose now $H_o = 0$. Denote by S the closure in $\mathfrak{g}$ of the set of all $H \in \mathfrak{a}'$ such that $\Phi_f(H) \neq 0$. If $0 \in {}^c S$, we can obviously find a neighborhood V of 0 in $\mathfrak{a}$, such that $\sup_{H \in V'} |\Phi_f(H)| < +\infty$. Let us assume $0 \in S$. Consider again the relation

$$(*) \qquad \det(t - \operatorname{ad} X) = t^n + p_1(X)t^{n-1} + \ldots + p_{n-\ell}(X)t^\ell \qquad (X \in \mathfrak{g}) .$$

The polynomials $p_1, \ldots, p_{n-\ell}$ are G-invariant and vanish for $X = 0$. Since $0 \in S \subset \mathfrak{a} \cap \mathrm{Cl}((\operatorname{Supp} f)^G)$, we have $\min_{X \in \operatorname{Supp} f} \sum_{1 \leq i \leq n-\ell} |p_i(X)| = 0$. Hence there exists an $X \in \operatorname{Supp} f$ such that $p_i(X) = 0$ for all i. This implies $X \in \mathcal{N}$ and contradicts the assumption $\mathcal{N} \cap \operatorname{Supp} f = \emptyset$. This completes the proof.

The proof of the theorem proceeds in the following way.

Let $\mathfrak{g}_o$ be the set of all $X_o \in \mathfrak{g}$ with the following property. There exists an open neighborhood ω of X_o in $\mathfrak{g}$ such that $\sup_{H \in \mathfrak{a}'} |\Phi_f(H)| < \infty$ for all $f \in C_c^\infty(\mathfrak{g})$ with $\operatorname{Supp} f \subset \omega$. Observe that $\mathfrak{g}_o$ is an invariant open subset of $\mathfrak{g}$. It is obviously sufficient for proving Theorem 13 to show that $\mathfrak{g}_o = \mathfrak{g}$.

By Lemma 29 we have ${}^c\mathcal{N} \subset \mathfrak{g}_o$, since $\mathcal{N}$ is closed in $\mathfrak{g}$. So it suffices to prove that $\mathcal{N} \subset \mathfrak{g}_o$. We again distinguish between two cases:

$X_o \neq 0$ and $X_o = 0$ $(X_o \in \mathcal{N})$.

First we give a few algebraic lemmas.

§3. Some algebraic lemmas on nilpotent elements.

Let $\mathcal{N}$ be (as before) the variety of nilpotent element of $\mathfrak{g}$.

Lemma 30. $\mathcal{N}$ is the union of a finite number of G-orbits (cf. J.P. Serre, Cohomologie galoisienne, Springer Lecture Notes 5 (1965), III-34, Cor. 2; the result stated there for unipotent elements in G, can be carried over, without difficulty, to the nilpotent elements of $\mathfrak{g}$).

Let σ be any G-orbit in $\mathfrak{g}$, $\sigma = X_o^G$ $(X_o \in \mathfrak{g})$. Define $\dim \sigma = \dim(\mathfrak{g}/\mathfrak{z}_{X_o})$ where $\mathfrak{z}_{X_o}$ = centralizer of X_o in $\mathfrak{g}$.

Lemma 31. Let σ be any G-orbit in $\mathcal{N}$. Then σ is open in its closure in $\mathfrak{g}$. If $\sigma' \neq \sigma$ is any orbit contained in Cl (σ), then $\dim \sigma' < \dim \sigma$. Let $\mathcal{N}_q$ be the union of all orbits of dimension $\leq q$. Then, if σ is of dimension q, σ is open in $\mathcal{N}_q$.

Proof. (Suggested by T. A. Springer) If σ is a G-orbit in $\mathcal{N}$, then the dimension of σ and the "algebraic" dimension of the algebraic orbit $A(\sigma)$, determined by σ (over $\overline{\Omega}$), coincide. Let σ' be an orbit in the closure of σ, $\sigma' \neq \sigma$. Then $A(\sigma')$ is in the Zariski-closure of $A(\sigma)$ and it is known that in this case $\dim A(\sigma') < \dim A(\sigma)$. Hence $\dim \sigma' < \dim \sigma$.

By Lemma 30 it now follows that $\mathcal{N}_p$ is closed for all $p \geq 0$. Let σ be a G-orbit in $\mathcal{N}$ of dimension q. Since $\sigma = \text{Cl}(\sigma) \cap {}^c\mathcal{N}_{q-1}$, σ is open in its closure. Furthermore, if σ' is any G-orbit in $\mathcal{N}_q$, $\sigma' \neq \sigma$, then $\sigma \cap \text{Cl}(\sigma') = \emptyset$ and $\text{Cl}(\sigma') \subset \mathcal{N}_q$. Hence, again by Lemma 30, σ is open in $\mathcal{N}_q$.

Corollary. All G-orbits in $\mathcal{N}$ are locally compact.

Lemma 32 (Jacobson-Morosow). Let $\mathfrak{g}$ be a semisimple Lie algebra over a field of characteristic zero. If $X_o \neq 0$ is any nilpotent element in $\mathfrak{g}$, we can choose two elements H_o, $Y_o \in \mathfrak{g}$ such that

$$[H_o, X_o] = 2X_o, \ [H_o, Y_o] = -2Y_o, \ [X_o, Y_o] = H_o$$

(cf. Harish-Chandra, Am. J. Math. 86 (1964), 271-309, Lemma 24).

The following lemma is well-known and is stated here only for reference.

Lemma 33. Let $\mathfrak{L}$ be a 3-dimensional Lie algebra over a field of characteristic zero, spanned by three elements H, X, Y with the relations

$$[H, X] = 2X, \ [H, Y] = -2Y, \ [X, Y] = H \ .$$

Then $\mathfrak{L}$ is semisimple. Let ρ be a representation of $\mathfrak{L}$ on a finite dimensional vector space V. Then 1) ρ is fully reducible; 2) $\rho(X)$, $\rho(Y)$ are nilpotent; 3) $\rho(H)$ is semisimple and all its eigenvalues are rational integers. Now suppose V is irreducible and dim V = m+1 $(m \geq 0)$. Then the eigenvalues of $\rho(H)$ are $m - 2r$ $(0 \leq r \leq m)$ and each of them has multiplicity 1. Let V_m be the subspace of V corresponding to the eigenvalue m. Then V_m is also the space of all $v \in V$ such that $\rho(X)v = 0$. Finally ρ is absolutely irreducible and apart from equivalence, there is exactly one irreducible representation of $\mathfrak{L}$ of degree d for any integer $d \geq 1$. (Cf. previous reference, Lemma 25.)

§4. A submersive map.

For any $X \in \mathfrak{g}$ put $\mathfrak{z}_X$ = centralizer of X in $\mathfrak{g}$ and $\mathfrak{g}_X = [X, \mathfrak{g}]$. Fix $X_o \in \mathcal{N}$, $X_o \neq 0$. Choose with Lemma 32, H_o, $Y_o \in \mathfrak{g}$ such that

$$[H_o, X_o] = 2X_o, \ [H_o, Y_o] = -2Y_o, \ [X_o, Y_o] = H_o \ .$$

Denote by $\mathfrak{L}$ the Lie subalgebra of $\mathfrak{g}$ spanned (over Ω) by X_o, Y_o, H_o. $\mathfrak{L}$ is simple. Let ρ = restriction of the adjoint representation of $\mathfrak{g}$ to $\mathfrak{L}$. Put $U = \mathfrak{z}_{Y_o}$ and $V = \mathfrak{g}_{Y_o}$. We have $U \cap \mathfrak{g}_{X_o} = \ker \rho(Y_o) \cap \operatorname{Im} \rho(X_o) = \{0\}$ by Lemma 33. For the same reasons $V \cap \mathfrak{z}_{X_o} = \operatorname{Im} \rho(Y_o) \cap \ker \rho(X_o) = \{0\}$.

Since $\dim \operatorname{Im} \rho(X_o) + \dim \ker \rho(Y_o) = \dim \ker \rho(X_o) + \dim \operatorname{Im} \rho(Y_o) = \dim \mathfrak{g}$ (again by Lemma 33), we get

$$U + \mathfrak{g}_{X_o} = V + \mathfrak{z}_{X_o} = \mathfrak{g} \quad \text{(direct sums)} .$$

The following mapping $\psi : G \times U \longrightarrow \mathfrak{g}$ plays an important role. It is defined by $\psi : (x, u) \longmapsto (X_o + u)^x$ where $x \in G$, $u \in U$. We investigate where ψ is submersive. Let us compute the differential of ψ at (x, u).

$$\frac{d}{dt}\psi(x \exp tX, u) = [X, X_o + u]^x \qquad (X \in \mathfrak{g})$$

$$\frac{d}{dt}\psi(x, u + tw) = w^x \qquad (w \in U) .$$

Now consider the map

$$(w, v) \longmapsto w^x + [v, X_o + u]^x \qquad (w \in U, v \in V) .$$

For $u \in U$, define $\phi_u : U \times V \longrightarrow \mathfrak{g}$ by

$$\phi_u(w, v) = w + [v, X_o + u] .$$

Observe that $\dim U + \dim V = \dim \ker \rho(Y_o) + \dim \operatorname{Im} \rho(Y_o) = \dim \mathfrak{g}$. To show that ψ is submersive at (x, u), it suffices therefore to prove that ϕ_u is bijective. We have

$$\phi_o(U \times V) = U + [X_o, V] = U + [X_o, \mathfrak{g}] = U + \mathfrak{g}_{X_o} = \mathfrak{g} .$$

So ϕ_o is a bijective linear map. Write

$$\phi_u(w, v) = \phi_o(w, v) + [v, u] \qquad (w \in U, v \in V) .$$

It follows that the mapping $u \longmapsto \phi_u \phi_o^{-1} - 1$ of U into $\operatorname{End}(\mathfrak{g})$ is linear. Put $Q(u) = \det(\phi_u \phi_o^{-1})$.

Q is a polynomial function, $Q(0) = 1$. So ϕ_u is bijective iff $Q(u) \neq 0$, and the set of these points u is open in U (cf. also Harish-Chandra, Amer.

J. Math. 86 (1964), 271-309, §7). We can actually prove that $Q = 1$ on U. We shall show this soon.

By a well-known theorem of Chevalley, $\mathfrak{L}$ is an algebraic Lie algebra (char $\Omega = 0$): there exists a connected algebraic subgroup $\underset{\sim}{L}$ of $\underset{\sim}{G}$, defined over Ω, such that $\mathfrak{L}$ is the Lie algebra of $L = \underset{\sim}{L} \cap G$. $\underset{\sim}{L}$ is semisimple. Let Γ be the (Cartan) subgroup of L with Lie algebra $\Omega . H_o$. Actually Γ is the centralizer of H_o in L. Let $\mathfrak{g} = \sum_{1 \leq i \leq r} \mathfrak{g}_i$ be a decomposition of $\mathfrak{g}$ into irreducible $\mathfrak{L}$-modules ($\mathfrak{L}$ acting on $\mathfrak{g}$ by means of ρ). Put $\dim \mathfrak{g}_i = \lambda_i + 1$ ($\lambda_i \geq 0$). So $\dim \mathfrak{g} = n = \sum_{1 \leq i \leq r} (\lambda_i + 1)$. We assume $\mathfrak{g}_1 = \mathfrak{L}$. So $\lambda_1 = 2$. It is not difficult to see that $U = \sum_{1 \leq i \leq r} U \cap \mathfrak{g}_i$ and $\dim(U \cap \mathfrak{g}_i) = 1$ (Lemma 33). Hence $r = \dim U$. Let ξ be the rational character of Γ, defined over Ω, given by the relation

$$\mathrm{Ad}(\gamma) Y_o = \xi(\gamma^{-1}) Y_o \qquad (\gamma \in \Gamma) .$$

Notice that $\xi \neq 1$. Furthermore $\mathrm{Ad}(\gamma) X_o = \xi(\gamma) X_o$ $(\gamma \in \Gamma)$. Using this, we easily see that we can choose the decomposition $\mathfrak{g} = \sum_{1 \leq i \leq p} \mathfrak{g}_i$ in irreducible $\mathfrak{L}$-modules, introduced above, in such a way that the following lemma holds.

Lemma 34. Let $r = \dim U$. We can choose a base $Y_o = w_1, w_2, \ldots, w_r$ for U such that $w_i \in U \cap \mathfrak{g}_i$ $(1 \leq i \leq r)$, and rational characters $\xi = \xi_1, \xi_2, \ldots, \xi_r$, defined over Ω, of Γ, such that

(i) $\xi_i^2 = \xi^{\lambda_i}$,

(ii) $\rho(H_o) w_i = -\lambda_i w_i$,

(iii) $\mathrm{Ad}(\gamma) w_i = \xi_i(\gamma^{-1}) w_i$, $(1 \leq i \leq r)$.

Concerning the proof, look at the highest and lowest weights of the irreducible representations in a decomposition of ρ.

In the above notations, put $\psi_u = \phi_u \phi_o^{-1}$ ($\in \mathrm{End}(\mathfrak{g})$) and let

$$u_\gamma = \xi(\gamma)\mathrm{Ad}(\gamma^{-1})u \qquad (\gamma \in \Gamma,\ u \in U) .$$

Since U is stable under $\mathrm{Ad}(\Gamma)$, we have $u_\gamma \in U$. Notice that V also is stable under $\mathrm{Ad}(\Gamma)$.

Lemma 35. $\psi_{u_\gamma} = \mathrm{Ad}(\gamma^{-1})\, \psi_u\, \mathrm{Ad}(\gamma)$ $(\gamma \in \Gamma,\ u \in U)$.

Proof. We have $\psi_{u_\gamma} = \phi_{u_\gamma} \phi_o^{-1}$; $\phi_u(w, v) = w + [X_o + u, v]$ $(w \in U,\ v \in V)$. Furthermore

$$\begin{aligned}\phi_{u_\gamma}(w, v) &= w + [X_o + u_\gamma, v] = w + \xi(\gamma)[\xi(\gamma)^{-1}X_o + u^{\gamma^{-1}}, v] \\ &= w + \xi(\gamma)[X_o + u)^{\gamma^{-1}}, v] = w + [X_o + u, \xi(\gamma)v^\gamma]^{\gamma^{-1}} \\ &= \mathrm{Ad}(\gamma^{-1})\{w^\gamma + [X_o + u, \xi(\gamma)v^\gamma]\} .\end{aligned}$$

We get $\mathrm{Ad}(\gamma)\phi_{u_\gamma}(w, v) = \phi_u(w^\gamma, \xi(\gamma)v^\gamma)$ $(w \in U,\ v \in V)$. Now take $u = 0$. Then $\mathrm{Ad}(\gamma)\phi_o(w, v) = \phi_o(w^\gamma, \xi(\gamma)v^\gamma)$ $(w \in U,\ v \in V)$. Summarizing we have $\mathrm{Ad}(\gamma)\phi_{u_\gamma} = \psi_u \mathrm{Ad}(\gamma)\phi_o$. So $\phi_{u_\gamma}\phi_o^{-1} = \mathrm{Ad}(\gamma^{-1})\psi_u \mathrm{Ad}(\gamma)$. This completes the proof.

Corollary. Put $Q(u) = \det(\psi_u)$ $(u \in U)$. Then $Q = 1$.

Proof. Choose $\gamma \in \Gamma$ such that $|\xi(\gamma)| < 1$. By Lemma 35 we have $Q(u_\gamma) = Q(u)$ $(u \in U)$. So $Q(u) = Q(\xi(\gamma)u^{\gamma^{-1}}) = Q(\xi(\gamma^m)u^{\gamma^{-m}})$ for all integers $m \geq 1$. Choose now a base $w_1, \ldots, w_r$ of U as in Lemma 34. Put $u = \sum_{1 \leq i \leq r} u_i w_i$ $(u_i \in \Omega)$. Then

$$u^{\gamma^{-m}} = \mathrm{Ad}(\gamma^{-m})u = \sum_{1 \leq i \leq r} u_i \mathrm{Ad}(\gamma^{-m})w_i = \sum_{1 \leq i \leq r} u_i \xi_i(\gamma^m)w_i$$

Hence $|\mathrm{Ad}(\gamma^{-m})u| \leq c_u \max_{1 \leq i \leq r} |\xi_i(\gamma^m)| = c_u \max_{1 \leq i \leq r} |\xi(\gamma)|^{\frac{m\lambda_i}{2}}$ by Lemma 34, where

c_u is a (real) constant, depending on u. So $|u^{\gamma^{-m}}| \le c_u$ for all m. Consequently $\xi(\gamma^m)u^{\gamma^{-m}} \longrightarrow 0$ $(m \longrightarrow \infty)$ and $Q(u) = Q(0) = 1$. This proves the lemma.

§5. Some more preparation.

The map $\psi : G \times U \longrightarrow \mathfrak{g}$, given by $\psi(x, u) = (X_o + u)^x$ $(x \in G,\ u \in U)$ is (everywhere) submersive. The image $\omega = \psi(G \times U)$ is an open and G-invariant subset of $\mathfrak{g}$. By Theorem 11 and its corollary, we have a linear map $\alpha \longmapsto f_\alpha$ of $C_c^\infty(G \times U)$ onto $C_c^\infty(\omega)$, such that

$$\int_{\mathfrak{g}} f_\alpha(X)F(X)dX = \int_{G\times U} \alpha(x, u)F((X_o + u)^x)dxdu$$

for every locally summable function F on ω.

In the notations of §4, put $t = \xi(\gamma)$ and write again $u_\gamma = \xi(\gamma)\mathrm{Ad}(\gamma^{-1})u$ for $u \in U$, $\gamma \in \Gamma$. We have $(X_o + u_\gamma)^{x\gamma} = (X_o + tu^{\gamma^{-1}})^{x\gamma} = t(X_o + u)^x$. Hence $t\omega = \omega$.

In the real case the differential operator "D" is the key to the solution (cf. Harish-Chandra, Amer. J. Math. 86 l. c.); in the $\mathfrak{p}$-adic case the stretching $X \longmapsto tX$ yields the desired result.

Lemma 36. Fix $\gamma \in \Gamma$ and $\alpha \in C_c^\infty(G \times U)$. Define $\alpha'(x, u) = \alpha(x\gamma^{-1}, u_{\gamma^{-1}})$ $(x \in G,\ u \in U)$. Then

$$f_\alpha(t^{-1}X) = |t|^{\frac{1}{2}(n-r)} f_{\alpha'}(X) \qquad (X \in \omega).$$

Proof. Let F be any function in $C_c^\infty(\omega)$. We have

$$\int_{\mathfrak{g}} f_\alpha(t^{-1}X)F(X)dX = \int_{\mathfrak{g}} f_\alpha(X)F(tX)|t|^n dX$$

$$= |t|^n \int_{G\times U} \alpha(x, u)F(t(X_o + u)^x)dxdu = |t|^n \int_{G\times U} \alpha(x, u)F((X_o + u_\gamma)^{x\gamma})dxdu$$

$$= |t|^n \int_{G\times U} \alpha(x\gamma^{-1}, u_{\gamma^{-1}})F((X_o + u)^x)\left|\frac{du_{\gamma^{-1}}}{du}\right|dxdu .$$

Let us compute the Jacobian $\left|\frac{du_{\gamma^{-1}}}{du}\right|$. Choose a base $w_1, \ldots, w_r$ of U as in Lemma 34. Put $u = \sum_{1\leq i\leq r} u_i w_i$ $(u_i \in \Omega)$. Then

$$u_{\gamma^{-1}} = t^{-1}\mathrm{Ad}(\gamma)u = t^{-1}\sum_{1\leq i\leq r} u_i \mathrm{Ad}(\gamma)w_i = t^{-1}\sum_{1\leq i\leq r} u_i \xi_i(\gamma^{-1})w_i$$

So u_i goes over in $t^{-1}\xi_i(\gamma)u_i$ $(1 \leq i \leq r)$. Hence

$$\left|\frac{du_{\gamma^{-1}}}{du}\right| = |t|^{-r}\prod_{1\leq i\leq r}|\xi_i(\gamma^{-1})| = |t|^{-r}\prod_{1\leq i\leq r}|t|^{-\frac{\lambda_i}{2}} = |t|^{-r-\frac{1}{2}\Sigma\lambda_i}$$

$$= |t|^{-r-\frac{1}{2}(n-r)} = |t|^{-\frac{1}{2}(n+r)}$$

Put $\alpha'(x, u) = \alpha(x\gamma^{-1}, u_{\gamma^{-1}})$ $(x \in G,\ u \in U)$. Then

$$\int_{\mathfrak{g}} f_\alpha(t^{-1}X)F(X)dX = |t|^{\frac{1}{2}(n-r)}\int_{G\times U}\alpha'(x, u)F((X_0 + u)^x)dudx$$

$$= |t|^{\frac{1}{2}(n-r)}\int_{\mathfrak{g}} f_{\alpha'}(X)F(X)dX \ .$$

Since F was arbitrary, we get $f_\alpha(t^{-1}X) = |t|^{\frac{1}{2}(n-r)} f_{\alpha'}(X)$ $(X \in \omega)$. This proves the lemma.

Fix $H \in \mathfrak{a}'$. Consider the distribution T_H on ω given by

$$T_H(f) = \Phi_f(H) = |\eta(H)|^{1/2}\int_{G/A} f(x^*H)dx^* \qquad (f \in C_c^\infty(\omega)) \ .$$

T_H is a G-invariant distribution: $T(f^y) = T(f)$ for all $f \in C_c^\infty(\omega)$, $y \in G$. With the same method of proof as in Lemma 21 we can show the existence of a unique distribution τ_H on U such that

$$T_H(f_\alpha) = \tau_H(\beta_\alpha) \text{ where } \beta_\alpha(u) = \int_G \alpha(x, u)dx \qquad (u \in U),\ \alpha \in C_c^\infty(G\times U) \ .$$

Recall ℓ = rank $\underset{\sim}{G}$. Let $f \in C_c^\infty(\omega)$ and put $f'(X) = f(t^{-1}X)$ $(X \in \omega)$. Then one easily checks the relation

$$T_H(f') = |t|^{\frac{1}{2}(n-\ell)} \Phi_f(t^{-1}H) .$$

Now fix $\alpha \in C_c^{\infty}(G \times U)$. Put

$f = f_\alpha$, $f'(X) = f(t^{-1}X) = |t|^{\frac{1}{2}(n-r)} f_{\alpha'}(X)$ $(X \in \omega)$, where $\alpha'(x, u) = \alpha(x\gamma^{-1}, u_{\gamma^{-1}})$

$(x \in G,\ u \in U)$, $\beta = \beta_\alpha$ and $\beta' = \beta_{\alpha'}$. We have

$$\beta'(u) = \int_G \alpha(x\gamma^{-1}, u_{\gamma^{-1}})dx = \beta(u_{\gamma^{-1}}) = \beta(t^{-1}u^{\gamma}) \qquad (u \in U) .$$

So we obtain

$$|t|^{\frac{1}{2}(n-\ell)} \Phi_f(t^{-1}H) = T_H(f') = |t|^{\frac{1}{2}(n-r)} T_H(f_{\alpha'}) = |t|^{\frac{1}{2}(n-r)} \tau_H(\beta') .$$

Or

$$\Phi_f(t^{-1}H) = |t|^{-\frac{1}{2}(r-\ell)} \tau_H(\beta') = |t|^{-\frac{1}{2}(r-\ell)} \{\Phi_f(H) + \tau_H(\beta' - \beta)\}$$

$(t = \xi(\gamma),\ H \in \mathfrak{a}')$.

It is well-known that $\dim U = \dim \mathfrak{z}_{X_0} = \dim \mathfrak{z}_{Y_0} = r \geq \ell$.

§6. The second step in the proof of Theorem 13.

Let us recall our aim. Let $X_0 \in \mathfrak{g}$ be nilpotent, $X_0 \neq 0$. We want to construct an open neighborhood ω_0 of X_0 such that Φ_f is bounded on $\mathfrak{a}'$ as soon as $\operatorname{Supp} f \subset \omega_0$ $(f \in C_c^{\infty}(\mathfrak{g}))$.

Let, as before, $\mathcal{N}_q$ = union of all G-orbits in $\mathcal{N}$ of dimension $\leq q$. Notice that $\mathcal{N}_{n-\ell} = \mathcal{N}$.

We have $X_0 \in \mathcal{N}_{n-r}$. By Lemma 31, X_0^G is open in $\mathcal{N}_{n-r}$. So we can choose an open neighborhood ω_1 of X_0 in ω such that $\omega_1 \cap \mathcal{N}_{n-r} \subset X_0^G$. We may assume $\omega_1 = \omega_1^G$.

Lemma 37. We can choose an open neighborhood U_1 of zero in U such that (i) $X_0 + U_1 \subset \omega_1$, (ii) $(X_0 + U_1) \cap X_0^G = \{X_0\}$.

The proof is exactly the same as in the real case (cf. Harish-Chandra, Am. J. Math. 86 (1964), 271-309, Lemma 22).

Fix $\gamma \in \Gamma$ such that $|t| = |\xi(\gamma)| > 1$. Choose an open neighborhood U_o of zero in U_1 such that $t^{-1}U_o^{\gamma} \cup tU_o^{\gamma^{-1}} \subset U_1$. Put $\mathcal{N}^* = \mathcal{N} - \{0\}$. Let $\mathfrak{g}_o$ be as in §2.

Lemma 38. $\mathcal{N}^* \subset \mathfrak{g}_o$.

Proof. The proof is by induction on $r = \dim \mathfrak{z}_{X_o}$ $(X_o \in \mathcal{N}^*)$. As observed above, we have $r \geq \ell$. In the above notations, put $(X_o + U_o)^G = \omega_o$. This is an open invariant neighborhood of X_o. Consider the restriction of the mapping $(x, u) \mapsto (X_o + u)^x$ to $G \times U_o$. Its image is ω_o. Let $\alpha \mapsto f_\alpha$ be the surjective linear map $C_c^\infty(G \times U_o) \longrightarrow C_c^\infty(\omega_o)$ given by Theorem 11 (which may of course be viewed as the restriction of the map $\alpha \mapsto f_\alpha$, considered in §5, to $C_c^\infty(G \times U_o)$). Let $f \in C_c^\infty(\omega_o)$ and choose $\alpha \in C_c^\infty(G \times U_o)$ such that $f = f_\alpha$. Define $\beta = \beta_\alpha$, β' and f' as before. We have $\beta \in C_c^\infty(U_o)$, $\beta' \in C_c^\infty(U_1)$, hence $\beta - \beta' \in C_c^\infty(U_1)$. Now $\beta'(u) = \beta(t^{-1}u^\gamma)$ $(u \in U_1)$, so $(\beta - \beta')(0) = 0$ and since $\beta - \beta'$ is locally constant, $0 \in {}^c\mathrm{Supp}(\beta - \beta')$. Define $\alpha_o(x, u) = \alpha_1(x)(\beta(u) - \beta'(u))$ $(x \in G,\ u \in U_1)$, where $\alpha_1 \in C_c^\infty(G)$ and $\int_G \alpha_1(x)dx = 1$. For $H \in \mathfrak{a}'$ we have

$$\Phi_{f_{\alpha_o}}(H) = T_H(f_{\alpha_o}) = \tau_H(\beta_{\alpha_o}) = \tau_H(\beta - \beta') .$$

We claim $\mathrm{Supp}\, f_{\alpha_o} \cap \mathcal{N}_{n-r} = \emptyset$.

By Theorem 11, $\mathrm{Supp}\, f_{\alpha_o} \subset \{X_o + \mathrm{Supp}(\beta - \beta')\}^G \subset (X_o + U_1)^G \subset \omega_1$. Hence $\mathrm{Supp}\, f_{\alpha_o} \cap \mathcal{N}_{n-r} \subset \omega_1 \cap \mathcal{N}_{n-r} \subset X_o^G$.

More precisely, if $x \in \mathrm{Supp}\, f_{\alpha_o} \cap \mathcal{N}_{n-r}$, then $x = (X_o + u)^{x'}$ for some $u \in \mathrm{Supp}(\beta - \beta')$, $x' \in G$. Hence $X_o + u \in X_o^G$ and $u \in U_1$. By Lemma 37: $u = 0$. But this contradicts $0 \in {}^c\mathrm{Supp}(\beta - \beta')$. Therefore $\mathrm{Supp}\, f_{\alpha_o} \cap \mathcal{N}_{n-r} = \emptyset$.

Let us now apply induction to $r = \dim \mathfrak{z}_{X_o}$.

(1) $r = \ell$ (the starting level). We have (cf. §5)

$$\Phi_f(t^{-1}H) = \Phi_f(H) + \tau_H(\beta' - \beta) \qquad (H \in \mathfrak{a}') .$$

Iteration gives

$$\Phi_f(t^{-m}H) = \Phi_f(H) + \sum_{1 \leq k \leq m} \tau_{t^{-k}H}(\beta' - \beta) \qquad (m \geq 1,\ H \in \mathfrak{a}') .$$

Fix a norm $|\cdot|$ on $\mathfrak{a}$ (e.g. the maximum of the coordinates w.r.t. some base of $\mathfrak{a}$).

Since $\operatorname{Supp} f_{a_o} \cap \mathcal{N} = \emptyset$ ($\mathcal{N} = \mathcal{N}_{n-\ell}$), we have $0 \notin \operatorname{Cl}\{$set of all $H \in \mathfrak{a}'$ with $\Phi_{f_{a_o}}(H) \neq 0\}$ (see e.g. proof of Lemma 29 (ii)). Therefore we can choose $\delta_1 > 0$ such that $\tau_H(\beta - \beta') = \Phi_{f_{a_o}}(H) = 0$ if $|H| < \delta_1$ $(H \in \mathfrak{a}')$. On the other hand we can choose (cf. corollary of Lemma 28) $\delta_2 > \delta_1$ such that $\tau_H(\beta - \beta') = 0$ if $|H| > \delta_2$ $(H \in \mathfrak{a}')$.

Applying again that $\operatorname{Supp} f_{a_o} \cap \mathcal{N} = \emptyset$, it follows from Lemma 28 that $\Phi_{f_{a_o}}$ is bounded on $\mathfrak{a}'$.

Put $C = \sup_{H \in \mathfrak{a}'} |\tau_H(\beta' - \beta)|$, $C_1 = \dfrac{C \log(\delta_2/\delta_1)}{\log |t|}$ and $C_2 = C + C_1$. We shall prove $|\Phi_f(H)| \leq C_2$ $(H \in \mathfrak{a}')$.

Two cases:

(i) $|H| \geq \delta_1$ $(H \in \mathfrak{a}')$.

Consider $\Phi_f(H) = \Phi_f(t^m H) + \sum_{1 \leq k \leq m} \tau_{t^k H}(\beta' - \beta)$ $(m \geq 1)$. Choosing m so large that $\Phi_f(t^m H) = 0$ ($H \in \mathfrak{a}'$ fixed), we may write

$$|\Phi_f(H)| \leq \sum_{1 \leq k < \infty} |\tau_{t^k H}(\beta' - \beta)| \qquad (H \in \mathfrak{a}') .$$

We get

$$|\Phi_f(H)| \leq C \frac{\log(\delta_2/\delta_1)}{\log |t|} = C_1 .$$

(ii) $|H| < \delta_1$ $(H \in \mathfrak{a}')$.

Let m be the smallest integer such that $|t^m H| \geq \delta_1$. Then

$$\Phi_f(H) = \Phi_f(t^m H) + \tau_{t^m H}(\beta' - \beta) .$$

So

$$|\Phi_f(H)| \leq C_1 + C = C_2 .$$

Summarizing: $|\Phi_f(H)| \leq C_2$ for all $H \in \mathfrak{a}'$.

(2) Now assume $r > \ell$. By the induction hypothesis and Lemma 29, $\operatorname{Supp} f_{a_o} \subset \mathfrak{g}_o$, hence $\Phi_{f_{a_o}}$ is bounded on $\mathfrak{a}'$. So again $\sup_{H \in \mathfrak{a}'} |\tau_H(\beta' - \beta)| = C < +\infty$. Put $c = |t|^{-\frac{1}{2}(r-\ell)} < 1$. We have

$$\Phi_f(t^{-m}H) = |t|^{-m(\frac{r-\ell}{2})} \Phi_f(H) + \sum_{1 \leq k \leq m} |t|^{-k(\frac{r-\ell}{2})} \tau_{t^{k-m}H}(\beta' - \beta) \quad (m \geq 1) .$$

Or

$$\Phi_f(H) = c^m \Phi_f(t^m H) + \sum_{1 \leq k \leq m} c^k \tau_{t^k H}(\beta' - \beta) \qquad (m \geq 1) .$$

Now $\lim_{m \to \infty} \Phi_f(t^m H) = 0$. So

$$|\Phi_f(H)| \leq C \sum_{1 \leq k < \infty} c^k \leq C \frac{c}{1 - c} \qquad (H \in \mathfrak{a}') .$$

Since $f \in C_c^\infty(\omega_o)$ was arbitrary (given $X_o \neq 0$), the lemma follows.

§7. Completion of the proof of Theorem 13.

To complete the proof of Theorem 13, it remains to show that $0 \in \mathfrak{g}_o$. This is easy if we apply again a stretching $X \mapsto tX$ $(X \in \mathfrak{g},\ t \in \Omega^*)$. Let $f \in C_c^\infty(\mathfrak{g})$. If $0 \notin \operatorname{Supp} f$, Φ_f is bounded on $\mathfrak{a}'$ by Lemma 29 and Lemma 38. So assume $0 \in \operatorname{Supp} f$. Fix $t \in \Omega^*$, $|t| > 1$. Put $f_t(X) = f(t^{-1}X)$ $(X \in \mathfrak{g})$. We have

$$\Phi_{f_t}(H) = |\eta(H)|^{1/2} \int_{G/A} f(t^{-1}x^* H)dx^* = |t|^{\frac{n-\ell}{2}} \Phi_f(t^{-1}H) \qquad (H \in \mathfrak{a}') .$$

So

$$\Phi_f(t^{-1}H) = |t|^{-(\frac{n-\ell}{2})} \Phi_{f_t}(H) = |t|^{-(\frac{n-\ell}{2})} \{\Phi_f(H) + \Phi_{f_t - f}(H)\} \qquad (H \in \mathfrak{a}') .$$

Since $0 \in \overset{c}{\mathrm{Supp}}(f_t - f)$, $C = \sup_{H \in \mathfrak{a}'} |\Phi_{f_t - f}(H)| < +\infty$. Put $c' = |t|^{-(\frac{n-\ell}{2})} < 1$. We get $|\Phi_f(t^{-1}H)| \leq c'\{|\Phi_f(H)| + C\}$ and by induction

$$|\Phi_f(t^{-m}H)| \leq c'^m |\Phi_f(H)| + (c' + c'^2 + \dots + c'^m)C$$

$$\leq |\Phi_f(H)| + \frac{c'}{1 - c'} \cdot C \qquad (m \geq 1) .$$

Put $C' = \frac{c'}{1 - c'} \cdot C$. We have

$$|\Phi_f(H)| \leq |\Phi_f(t^m H)| + C' \qquad (m \geq 1,\ H \in \mathfrak{a}') ,$$

hence

$$|\Phi_f(H)| \leq C' \quad \text{for all } H \in \mathfrak{a}' .$$

This completes the proof of Theorem 13.

§8. Lifting of Theorem 13 to the group.

We come now to the proof of the "boundedness" of F_h. We shall prove the following theorem.

Theorem 14. *Let* A *be any Cartan subgroup of* G *and let* ω_A *be a compact subset of* A. *Put* $\omega_A' = \omega_A \cap G'$. *Then*

$$\sup_{a \in \omega_A'} |F_h(a)| < +\infty$$

for all $h \in C^\infty(G)$ *with compact support* mod Z.

The proof of this theorem is very similar to (Harish-Chandra, A formula for semisimple Lie groups, Am. J. Math. 79 (1957), 733-760, Theorem 2).

Lemma 39. Let $\omega \subset G$ be compact mod Z. Then the set of all $a \in A$ satisfying $a \in Cl(\omega^G)$ is relatively compact mod Z in G.

The proof is similar to its infinitesimal analogue: Lemma 28. (There is no reason in this case to reduce the lemma to semisimple groups.) In order to prove Theorem 14, we make use of the following two lemmas.

Lemma 40. There exists an open neighborhood V of 1 in A with the following property. Put $V' = V \cap A'$. Then

$$\sup_{a \in V'} |F_h(a)| < +\infty$$

for all locally constant functions h with compact support mod Z.

Proof. With the notations of §1, write $\underline{G} = \underline{T}.\underline{G}_1$, $T = \underline{T} \cap G$, $G_1 = \underline{G}_1 \cap G$ and $G_o = T.G_1$. We recall that T is the group of Ω-rational points of the identity component of the center of $\underline{G}$, which is a (central) Ω-torus. The group G_o is an open subgroup of G of finite index. Put $A_o = A \cap G_o$, $A_1 = A \cap G_1$. Then A/A_o is finite and A_1 is a Cartan subgroup of G_1. Now observe that a priori V may be required to be contained in A_o and even $V \subset \bigcap_{y \in G_o \backslash G/A} (A^y \cap G_o)$. Then applying similar methods as in the proof of Lemma 26 and Lemma 27, it follows that we are reduced to the case where G is semisimple. The fact that Supp(h) is not compact does not harm. Notice that $C_c^\infty(G_o) \subset C_c^\infty(T) \otimes C_c^\infty(G_1)$.

From this point on, the proof is very similar to the real case. We have to find an open neighborhood V of 0 in $\mathfrak{g}$ such that

(i) "exp" is defined on V and $V^x = V$ for all $x \in G$ (we use the formula $\exp H^x = (\exp H)^x$ for defining "exp" on V).

(ii) "exp" is injective and submersive on V.

(iii) exp V is completely invariant: for any compact subset $\omega \subset \exp V$, $Cl(\omega^G) \subset \exp V$.

This being done, it is easy to put the function F_h over to the Lie algebra $\mathfrak{a}$ of A in a sufficiently small neighborhood of 1 in A, contained in exp V (cf. also Part VII, §1). We obtain of course the function Φ_f, where f is given by $f(H) = h(\exp H)$ $(H \in V)$. Now we can apply Theorem 13.

We refer to (Harish-Chandra, Am. J. Math. 79 (1957), 733-760, Lemma 15 and Trans. A.M.S. 119 (1965), 457-508, §3). We leave the details to the reader.

Lemma 41. Given $a_o \in A$, there exists a neighborhood V of a_o in A with the following properties. Put $V' = V \cap A'$. Then

$$\sup_{a \in V'} |F_h(a)| < +\infty$$

for all locally constant functions h with compact support mod Z.

Proof. Let Z_G be the center of G. If $a_o \in Z_G$, we have for any $h \in C^\infty(G)$ with compact support mod Z,

$$F_h(a_o a) = |D(a_o a)|^{1/2} \int_{G/A} h((aa_o)^{x^*})dx^*$$
$$= |D(a)|^{1/2} \int_{G/A} h((a)^{x^*} a_o)dx^* \qquad (a \in A') .$$

Hence the lemma reduces to the case $a_o = 1$, hence to Lemma 40. Now assume $a_o \in {}^cZ_G$. Fix $h \in C^\infty(G)$, Supp h compact mod Z. Let $\underset{\sim}{\Xi}$ and $\mathfrak{z}_o$ be the centralizers of a_o in $\underset{\sim}{G}$ and $\mathfrak{g}$ respectively. $(\underset{\sim}{\Xi})^o$ is a connected, reductive Ω-subgroup of G. Put $\Xi_o = (\underset{\sim}{\Xi})^o \cap G$. The Lie algebra of Ξ_o is $\mathfrak{z}_o$. We have $A \subset \Xi_o$. Therefore A is a Cartan subgroup of Ξ_o. For any $y \in \Xi_o$, put $\nu(y) = \det(1 - \mathrm{Ad}(y))_{\mathfrak{g}/\mathfrak{z}_o}$. We have $D(y) = D_o(y).\nu(y)$ $(y \in \Xi_o)$, where D_o is defined by $D_o(y) = \det(1 - \mathrm{Ad}(y))_{\mathfrak{z}_o}$ $(y \in \Xi_o)$.

Now $\nu(a_o) \neq 0$. Choose a neighborhood W of 1 in Ξ_o such that $|\nu(a_o y)| = |\nu(a_o)| \neq 0$ for $y \in W$.

Now choose with Lemma 19 a neighborhood V of 1 in A, $V \subset W$, such that $(a_o V)^x \cap \operatorname{Supp} h = \emptyset$ unless $\bar{x} \in \bar{C}$, $\bar{C}$ being a compact subset of G/Ξ_o.

Fix Haar measures dx, da, $d\xi$ on G, A, Ξ_o respectively. Choose invariant measures $d\xi^*$, dx^*, $d\bar{x}$ on Ξ_o/A, G/A and G/Ξ_o respectively such that $d\xi = d\xi^* da$, $dx = dx^* da$, $dx = d\bar{x} d\xi$. We have

$$\int_{G/A} h(a^{x^*})dx^* = \int_{G/\Xi_o} d\bar{x} \int_{\Xi_o/A} h((a^{\xi^*})^x)d\xi^* \qquad (a \in A') .$$

Now take $a \in (a_o V) \cap A'$. Then

$$\bar{x} \mapsto \int_{\Xi_o/A} h((a^{\xi^*})^x)d\xi^*$$

vanishes outside $\bar{C}$, chosen above. Choose $\alpha \in C_c^\infty(G)$ such that

$$\bar{\alpha}(\bar{x}) = \int_{\Xi_o} \alpha(x\xi)d\xi = \begin{cases} 1 & \text{if } \bar{x} \in \bar{C} \\ 0 & \text{elsewhere} \end{cases} \qquad (\bar{x} \in G/\Xi_o) .$$

Put $g(y) = \int_G \alpha(x)h((a_o y)^x)dx \quad (y \in \Xi_o)$. Then clearly $g \in C^\infty(\Xi_o)$, Supp g is compact mod Z. Furthermore, if $a \in a_o^{-1}((a_o V) \cap A')$, then

$$\int_{\Xi_o/A} g(a^{\xi^*})d\xi^* = \int_G \alpha(x)dx \int_{\Xi_o/A} h(((a_o a)^{\xi^*})^x d\xi^*$$

$$= \int_{G/\Xi_o} \bar{\alpha}(\bar{x})d\bar{x} \int_{\Xi_o/A} h(((a_o a)^{\xi^*})^x)d\xi^* = \int_{G/A} h((a_o a)^{x^*})dx^* .$$

Now observe that $a_o^{-1}((a_o V) \cap A')$ consists of those elements $a \in V$ for which $D_o(a) \neq 0$. We obtain

$$F_h(a_o a) = |D_o(a)|^{1/2} |\nu(a_o)|^{1/2} \int_{\Xi_o/A} g(a^{\xi^*})d\xi^* \qquad (a \in V \cap \Xi_o')$$

The result now follows from Lemma 40, applied to Ξ_o. This completes the proof.

Part VII. The local summability of $|D|^{-\frac{1}{2}-\varepsilon}$ (char $\Omega = 0$).

§1. Statement of Theorem 15. Reduction to the Lie algebra.

The main result of this Part is the following theorem.

Theorem 15. Given G, there exists $\varepsilon > 0$ such that the function $x \longmapsto |D(x)|^{-\frac{1}{2}-\varepsilon}$ is locally summable with respect to the Haar measure on G.

We assume again char $\Omega = 0$. The theorem implies of course the local summability of $x \longmapsto |D(x)|^{-\frac{1}{2}}$ ($x \in G$) (cf. Harish-Chandra, Trans. A.M.S. 119 (1965), 457-508, Lemma 53 for the real case). But, for later purposes (see next Part), we need more.

We shall reduce the proof of the theorem to the proof of a similar statement about Cartan subgroups and subsequently about Cartan subalgebras.

Let A be a Cartan subgroup of G with Lie algebra $\mathfrak{a} \subset \mathfrak{g}$. All Lie algebras will be assumed to be linear: if $G \subset GL(n, \Omega)$, we assume $\mathfrak{g} \subset \mathfrak{gl}(n, \Omega)$. Put $G_A = (A')^G$ as usual. Let dx, da denote the Haar measures on G and A respectively. Let $\tilde{A}$ be the normalizer of A in G and define $W_A = \tilde{A}/A$. W_A is a finite group with $[W_A]$ elements.

Lemma 42. Let dx^* be the invariant measure on $G^* = G/A$ such that $dx = dx^* da$. Then $\int_{G_A} f(x)dx = [W_A]^{-1} \int_A |D(a)| da \int_{G/A} f(a^{x^*}) dx^*$ ($f \in C_c(G_A)$).

The proof is exactly the same as in the real case (cf. Harish-Chandra, Acta Math. 116 (1966), 1-111, Lemma 91).

The group $\underset{\sim}{H} = \mathrm{Ad}(\underset{\sim}{G}) \subset GL(\mathfrak{g})$, where $\mathfrak{g}$ is the Lie algebra of $\underset{\sim}{G}$, is an Ω-group and Ad is an Ω-morphism. The Lie algebra of $\mathrm{Ad}(\underset{\sim}{G})$ is $\mathrm{ad}(\mathfrak{g})$. Since Ad is submersive at the neutral element of G, $\mathrm{Ad}(G)$ is an open subgroup of $H = \underset{\sim}{H} \cap GL(\mathfrak{g})$. Observe that $D(xz) = D(x)$ for $x \in \underset{\sim}{G}$,

$z \in \operatorname{Ker} Ad = Z_{\underset{\sim}{G}}$, the center of $\underset{\sim}{G}$. Moreover, $D_H(Ad\, x) = D(x)$ for $x \in \underset{\sim}{G}$, where D_H is the "D"-function of $\underset{\sim}{H}$. The formula

$$\int_G |D(x)|^{-\frac{1}{2}-\varepsilon} f(x)dx = \int_{Ad(G)} |D_H(x)|^{-\frac{1}{2}-\varepsilon} \int_{Z_G} f(xz)dz \qquad (f \in C_c(G))$$

shows that it is certainly enough to prove the local summability of D_H on H. Therefore, we are reduced to the case of semisimple groups. So assume, for the time being, G semisimple.

Up to conjugacy there are only finitely many Cartan subgroups: $A_1, \ldots, A_s$ (char $\Omega = 0$!). Fix $\varepsilon \geq 0$. Let $f \in C_c^{\infty}(G)$ be arbitrary, $f \geq 0$. We have (using integral signs for upper integral signs, in the terminology of Bourbaki)

$$\int_G |D(x)|^{-\frac{1}{2}-\varepsilon} f(x)dx = \sum_{1\leq i\leq s} \int_{G_{A_i}} |D(x)|^{-\frac{1}{2}-\varepsilon} f(x)dx$$

$$= \sum_{1\leq i\leq s} [W_{A_i}]^{-1} \int_{A_i} |D(a)|da \int_{G/A_i} |D(a)|^{-\frac{1}{2}-\varepsilon} f(a^{x^*})dx^*$$

$$= \sum_{1\leq i\leq s} [W_{A_i}]^{-1} \int_{A_i} |D(a)|^{\frac{1}{2}-\varepsilon} \int_{G/A_i} f(a^{x^*})dx^* .$$

Fix i $(1 \leq i \leq s)$ and put $A = A_i$. We get

$$\int_A |D(a)|^{\frac{1}{2}-\varepsilon} \int_{G/A} f(a^{x^*})dx^* = \int_A |D(a)|^{-\varepsilon} F_f(a)da$$

in the notation of Part VI. Since F_f is bounded on A' and vanishes outside some compact subset of A (cf. Lemma 39), it remains to prove:

Lemma 43. <u>Let</u> A <u>be a Cartan subgroup of</u> G. <u>There exists</u> $\varepsilon > 0$ <u>such that</u> $a \longmapsto |D(a)|^{-\varepsilon}$ <u>is locally summable on</u> A.

Observe that in case $\varepsilon = 0$ we obtain:

$$\int_G |D(x)|^{-\frac{1}{2}} f(x)dx = \sum_{1\le i\le s} [W_{A_i}] \int_{A_i} F_f(a)da < c \quad ,$$

where c is a (real) constant. So it follows now that $|D|^{-\frac{1}{2}}$ is locally summable on G. However, with Lemma 43 we prove more.

Let us reduce Lemma 43 to a similar statement about the Lie algebra $\mathfrak{a}$ of A. From now on we shall drop the assumption that G is semisimple. Let U and ω be open neighborhoods of 1 and 0 in A and $\mathfrak{a}$ respectively such that "exp" is defined on ω and maps ω isomorphically and homeomorphically onto U. It is obviously sufficient to investigate the existence of integrals of the form $\int_{a_o U_o} |D(a)|^{-\varepsilon} da$, where $a_o \in A$ and U_o is a (arbitrarily small) compact neighborhood of 1, contained in U. If a_o is regular, we may choose U_o such that $|D(a)| = |D(a_o)| \neq 0$ for $a \in a_o U_o$. The integral $\int_{a_o U_o} |D(a)|^{-\varepsilon} da$ exists. Now suppose a_o singular. $\underset{\sim}{A}$ (and $\underset{\sim}{\mathfrak{a}}$) splits over a finite field extension L of Ω. Let L be endowed with the unique valuation which extends that of Ω. Denote by Σ the set of roots of $(\mathfrak{g} \otimes L, \mathfrak{a} \otimes L)$. For $\alpha \in \Sigma$ let ξ_α be the corresponding L-rational character of $\underset{\sim}{A}$: if $\operatorname{ad} H(X_\alpha) = \alpha(H)X_\alpha$, then $\operatorname{Ad}(a)X_\alpha = \xi_\alpha(a)X_\alpha$ $(H \in \mathfrak{a},\ a \in A,\ \alpha \in \Sigma,\ X_\alpha \in \mathfrak{g} \otimes L)$. Write $\mathfrak{g}_L = \mathfrak{g} \otimes L$, $\mathfrak{a}_L = \mathfrak{a} \otimes L$. Let us compare $|D(a)|$ with $|D(a_o a)|$ $(a \in A)$. We have

$$|D(a_o a)| = |\det(1 - \operatorname{Ad}(a_o)\operatorname{Ad}(a))_{\mathfrak{g}_L/\mathfrak{a}_L}| = \prod_{\alpha\in\Sigma} |1 - \xi_\alpha(a_o)\xi_\alpha(a)|_L$$

$$= \prod_{\alpha\in\Sigma_o} |1 - \xi_\alpha(a)|_L \cdot \prod_{\alpha\in\Sigma\setminus\Sigma_o} |1 - \xi_\alpha(a_o)\xi_\alpha(a)|_L \ ,$$

where Σ_o is the set of roots $\alpha \in \Sigma$ such that $\xi_\alpha(a_o) = 1$.

Put

$$\nu(a) = \prod_{\alpha\in\Sigma\setminus\Sigma_o} |1 - \xi_\alpha(a_o)\xi_\alpha(a)|_L \ .$$

We choose U_o such that $\nu(a) = \nu(1) \neq 0$ for $a \in U_o$. So

$$|D(a_o a)| = \nu(1) \prod_{\alpha\in\Sigma_o} |1 - \xi_\alpha(a)|_L \qquad (a \in U_o) .$$

There exists a constant $\beta > 0$ such that $\sup_{a\in U_o} \prod_{\alpha\in\Sigma\setminus\Sigma_o} |1 - \xi_\alpha(a)|_L < \beta$. Hence

$$\beta . |D(a_o a)| = \nu(1) \prod_{\alpha\in\Sigma_o} |1 - \xi_\alpha(a)|_L . \beta \geq \nu(1)|D(a)| \qquad (a \in U_o) .$$

So

$$\int_{a_o U_o} |D(a)|^{-\varepsilon} da = \int_{U_o} |D(a_o a)|^{-\varepsilon} da \leq \left(\frac{\nu(1)}{\beta}\right)^{-\varepsilon} \int_{U_o} |D(a)|^{-\varepsilon} da .$$

Put $a = \exp H$ $(a \in U,\ H \in \omega)$. As in Part VI, §8 we have

$$D(\exp H) = \det(1 - \mathrm{Ad}(\exp H))_{\mathfrak{g}_L/\mathfrak{a}_L} = \det(1 - \exp \mathrm{ad}\, H)_{\mathfrak{g}_L/\mathfrak{a}_L} .$$

So

$$D(\exp H) = \prod_{\alpha\in\Sigma} (1 - e^{\alpha(H)}) .$$

On the other hand $\eta(H) = \det(\mathrm{ad}\, H)_{\mathfrak{g}_L/\mathfrak{a}_L} = \prod_{\alpha\in\Sigma} \alpha(H)$. Choose $\omega_o \subset \omega$ such that $U_o = \exp \omega_o$. We may assume that ω_o and U_o are compact and are so small that

$$\frac{|D(\exp H)|}{|\eta(H)|} = \prod_{\alpha\in\Sigma} \left|1 + \frac{\alpha(H)}{2} + \dots\right|_L = 1 \text{ for all } H \in \omega_o .$$

Now

$$\int_{U_o} |D(a)|^{-\varepsilon} da = \int_{\omega_o} |D(\exp H)|^{-\varepsilon} dH = \int_{\omega_o} |\eta(H)|^{-\varepsilon} dH .$$

Therefore, it suffices to prove the following lemma.

Lemma 44. <u>Let $\mathfrak{a}$ be a Cartan subalgebra of $\mathfrak{g}$. There exists $\varepsilon > 0$ such that $H \longmapsto |\eta(H)|^{-\varepsilon}$ is locally summable on $\mathfrak{a}$.</u>

§2. Proof of the main lemma.

We prove here Lemma 44. Put

$$T_s(f) = \int_{\mathfrak{a}} f(H)\,|\eta(H)|^s dH \qquad (f \in C_c^{\infty}(\mathfrak{a}),\ \mathcal{R}\, s > 0)\ .$$

Evidently $s \mapsto T_s(f)$ is holomorphic if $\mathcal{R}\, s > 0$. We may also write

$$T_s(f) = \sum_{-\infty<k<\infty} c_f(k) q^{-ks} \dagger$$

where

$$c_f(k) = \int_{|\eta(H)|=q^{-k}} f(H)dH\ .$$

Put $z = q^{-s}$. Thus $T_s(f) = \sum_{-\infty<k<\infty} c_f(k) z^k$. Notice that $c_f(k) = 0$ for k large and negative. The series

$$\Sigma\, |c_f(k)|\, |z|^k$$

converges for $|z| < 1$.

To prove the lemma, it suffices to show that this series converges for some z with $|z| > 1$. Hence it is enough to prove that $s \mapsto T_s(f)$ extends to a holomorphic function for $\mathcal{R}\, s > -\varepsilon$ for some $\varepsilon > 0$, independent of f.

We use induction on $\dim \mathfrak{g}$. By an easy argument (cf. proof of Lemma 27) we are reduced to the case where $\mathfrak{g}$ is semisimple. For $H_o \in \mathfrak{a}$, let $\mathfrak{z}_{H_o}$ denote the centralizer of H_o in $\mathfrak{g}$. In the notations of §1, $\mathfrak{z}_{H_o} \otimes L = \mathfrak{a}_L + \sum_{\alpha\in\Sigma(H_o)} LX_{\alpha}$, where $\Sigma(H_o)$ = set of all $\alpha \in \Sigma$ such that $\alpha(H_o) = 0$. We conclude that there are only finitely many possibilities for $\mathfrak{z}_{H_o}$.

If $H_o \neq 0$, we have $\dim \mathfrak{z}_{H_o} < \dim \mathfrak{g}$ and $\mathfrak{a} \subset \mathfrak{z}_{H_o}$. Since $\mathfrak{z}_{H_o}$ is reductive, the induction hypothesis applies to $\mathfrak{z}_{H_o}$.

Near H_o we have $\eta(H) = \eta_{\mathfrak{z}_{H_o}}(H).\nu(H)$, $\nu(H) = \prod_{\alpha\in{}^c\Sigma(H_o)} \alpha(H)$, $|\nu(H)| = |\nu(H_o)| \neq 0$. So, if $f \in C_c^{\infty}(\mathfrak{a})$ and $0 \in {}^c\mathrm{Supp}\, f$, it follows that

---

$\dagger$ q is the order of the residue class field of Ω.

$s \longmapsto T_s(f)$ is holomorphic for $\mathcal{R}\, s > -\varepsilon$ for some $\varepsilon > 0$, independent of f.

Now suppose $0 \in \operatorname{Supp} f$. Fix $t \in \Omega^*$, $|t| \neq 1$. Put

$$f_t(H) = f(t^{-1}H) \qquad (H \in \mathfrak{a}) .$$

Then

$$T_s(f_t) = \int_{\mathfrak{a}} f(t^{-1}H)\,|\eta(H)|^s dH = \int_{\mathfrak{a}} f(H)\,|\eta(tH)|^s d(tH)$$
$$= |t|^{\ell+s(n-\ell)} \int_{\mathfrak{a}} f(H)\,|\eta(H)|^s dH = |t|^{\ell+s(n-\ell)} T_s(f) .$$

Here $\ell = \operatorname{rank} \underset{\sim}{G}$, $n = \dim \underset{\sim}{G}$ (as usual). Therefore,

$$T_s(f_t - f) = (|t|^{\ell+s(n-\ell)} - 1) T_s(f) ,$$

and hence

$$T_s(f) = \frac{T_s(f_t - f)}{|t|^{\ell+s(n-\ell)} - 1} .$$

The numerator on the right-hand side is holomorphic if $\mathcal{R}\, s > -\varepsilon$, since $0 \notin \operatorname{Supp}(f_t - f)$. Also the denominator cannot vanish if

$$\mathcal{R}\, s > -\frac{\ell}{n-\ell} .$$

So if ε is sufficiently small, $s \longmapsto T_s(f)$ is holomorphic for $\mathcal{R}\, s > -\varepsilon$ independent of the choice of f. This completes the proof of the lemma.

Part VIII. The local summability of the characters of the supercuspidal representations (char $\Omega = 0$).

§1. The main theorem and its consequences.

We assume in this Part again: char $\Omega = 0$. Our main theorem is as follows.

Theorem 16. *Let* Θ *be the character of a supercuspidal representation of* G *and* F *the locally constant function on* G' *such that* $\Theta = F$ *on* G'. *Then* F *is locally summable on* G *and* $\Theta = F$.

Before proceeding with the proof of the main theorem, which requires considerable preparation, we derive some consequences.

We call a Cartan subgroup Γ of G *elliptic* if Γ/Z is compact.[1)]

Lemma 45 (Selberg principle). *Let* Γ *be any Cartan subgroup of* G. *Let* θ *be a supercusp form. Then*

$$\int_{G/\Gamma} \theta(\gamma^{\bar{x}})d\bar{x} = 0 \qquad (\gamma \in \Gamma')$$

unless Γ *is elliptic.*

Proof. Fix Γ and let A be the maximal split torus in Γ. Choose a parabolic subgroup P with A as a split component and let $P = MN = NM$ be the corresponding Levi-decomposition. Fix a compact subgroup K as in the theorem of Bruhat and Tits. We have $G = KP = KNM$. In a suitable normalization of the Haar measures

$$\int_{G/\Gamma} \theta(x\gamma x^{-1})d\bar{x} = \int_K \int_{M/\Gamma} \int_N \theta^k(nm\gamma m^{-1}n^{-1})dndmdk \qquad (\gamma \in \Gamma') .$$

By Lemma 22 we get

---

1) Such subgroups Γ always exist (cf. M. Kneser, Galois-Kohomologie halbeinfacher algebraischer Gruppen über $\mathfrak{p}$-adischen Körpern II, Math. Zeitschrift 89 (1965), 250-272, §15.

$$\int_N \theta^k(nm\gamma m^{-1}n^{-1})dn = |\det(\mathrm{Ad}(\gamma^{-1}) - 1)_{\mathfrak{n}}|^{-1}\int_N \theta^k(m\gamma m^{-1}n)dn$$

where $\mathfrak{n}$ is the Lie algebra of N, $\gamma \in \Gamma'$.

If Γ is not elliptic, we have $P \neq G$. Hence $\int_{G/\Gamma} \theta(\gamma^{\bar{x}})d\bar{x} = 0$ for $\gamma \in \Gamma'$. This proves the lemma.

Let $B_1, B_2, \ldots, B_r$ be a maximal set of non-conjugate elliptic Cartan subgroups of G. As usual, put $G_{B_i} = (B_i')^G$ $(1 \leq i \leq r)$. Define

$$G_e = \bigcup_{1\leq i\leq r} G_{B_i} .$$

Then G_e is an open subset of G. We call it the <u>elliptic set</u>. We normalize the Haar measures $d_i b^*$ on B_i/Z in such a way that

$$\int_{B_i/Z} d_i b^* = 1 \qquad (1 \leq i \leq r) .$$

Notice that in the real case r is at most 1. In the $\mathfrak{p}$-adic case r can be larger than 1 as already the group $G = GL(2)$ shows, where $r = 3$. For $\omega \in {}^0\mathcal{E}(G)$, denote by Θ_ω the character of ω. Fix $\chi \in \hat{Z}$. For $\omega \in {}^0\mathcal{E}(G, \chi)$ put

$$\Phi_{\omega, B_i}(b) = |D(b)|^{1/2}\Theta_\omega(b) \qquad (b \in B_i';\ 1 \leq i \leq r) .$$

Theorem 17. <u>Let</u> $\omega_1, \omega_2 \in {}^0\mathcal{E}(G, \chi)$. <u>Then</u>

$$\sum_{1\leq i\leq r} [W_{B_i}]^{-1} \int_{B_i/Z} \overline{\Phi_{\omega_1, B_i}(b)}\, \Phi_{\omega_2, B_i}(b)d_i b^* = \begin{cases} 0 & \text{if } \omega_1 \neq \omega_2 \\ 1 & \text{if } \omega_1 = \omega_2 \end{cases} .$$

<u>Proof</u>. Let $\omega \in {}^0\mathcal{E}(G, \chi)$. Fix $\pi \in \omega$. Choose a K-finite unit vector ϕ in the space of π and put

$$\theta(x) = (\phi, \pi(x)\phi) \qquad (x \in G) .$$

Then θ is a supercusp form. Let K_o be an open compact subgroup of G, whose Haar measure is normalized such that the total measure of K_o is 1. For any Cartan subgroup Γ of G we have by Theorem 12,

$$\Theta_\omega(\gamma) = d(\omega) \int_{G/Z} dx^* \int_{K_o} \theta(\gamma^{xk})dk \qquad (\gamma \in \Gamma') .$$

Now assume Γ elliptic. Then we are allowed to interchange the integrals, since (cf. the corollary of Lemma 19),

$$(x^*, k) \longmapsto \theta(\gamma^{xk})$$

has compact support in $G/Z \times K_o$ $(\gamma \in \Gamma')$. We get

$$\Theta_\omega(\gamma) = d(\omega) \int_{G/Z} \theta(\gamma^x)dx^* \qquad (\gamma \in \Gamma') .$$

So

$$\Phi_{\omega, B_i}(b) = d(\omega)|D(b)|^{1/2} \int_{G/Z} \theta(b^x)dx^* \qquad (b \in B_i', \ 1 \leq i \leq r) .$$

Put

$$Q = \int_{G/Z} \overline{\theta}(x)\pi(x)dx^* .$$

Then

$$\operatorname{tr} Q = \int_{G/Z} |\theta(x)|^2 dx^* = d(\omega)^{-1} .$$

On the other hand

$$\operatorname{tr} Q = \int_{G/Z} \Theta_\omega(x)\overline{\theta}(x)dx^* .$$

Hence, applying the Selberg principle,

$$\begin{aligned} \operatorname{tr} Q &= \sum_{1 \leq i \leq r} [W_{B_i}]^{-1} \int_{B_i/Z} \Phi_{\omega, B_i}(b)|D(b)|^{1/2} d_i b^* \int_{G/Z} \overline{\theta}(b^x)dx^* \\ &= \sum_{1 \leq i \leq r} [W_{B_i}]^{-1} d(\omega)^{-1} \int_{B_i/Z} |\Phi_{\omega, B_i}(b)|^2 d_i b^* \end{aligned}$$

(cf. Lemma 42). Hence

$$\sum_{1 \leq i \leq r} [W_{B_i}]^{-1} \int_{B_i/Z} |\Phi_{\omega, B_i}(b)|^2 d_i b^* = 1 .$$

The remaining part of the theorem is proved in the same way by considering, in place of Q, the operator $\int_{G/Z} \overline{\theta}_2(x)\pi_1(x)dx^*$, whose meaning is evident. This proves the theorem.

Corollary. Fix $\chi \in \hat{Z}$ and let

$$\Theta = c_1 \Theta_{\omega_1} + c_2 \Theta_{\omega_2} + \dots + c_n \Theta_{\omega_n}$$

where c_i are complex numbers and $\omega_1, \dots, \omega_n \in {}^0\mathcal{E}(G, \chi)$. If $\Theta = 0$ on G_e, then $\Theta = 0$.

This is an immediate consequence of Theorem 17.

§2. Statement of the preparatory results for the proof of Theorem 16.

Let $M(n, \Omega)$ denote the algebra of all $n \times n$ matrices with coefficients in Ω. Put

$$|x| = \max_{i,j} |x_{ij}| \qquad (x \in M(n, \Omega)) .$$

Then $|x + y| \leq \max(|x|, |y|)$ and $|xy| \leq |x| \cdot |y|$. For $x \in GL(n, \Omega)$, define

$$\|x\| = \max(|x|, |x|^{-1}) .$$

Then $\|xy\| \leq \|x\| \cdot \|y\|$ and $\|x\| \geq 1$ $(x \in GL(n, \Omega))$. Note that $\|x\| = 1$ iff $x \in GL(n, \mathcal{O})$, where $\mathcal{O}$ is the ring of integers in Ω. Let q be the number of elements of the residue class field of Ω. Define the function $\sigma : GL(n, \Omega) \longrightarrow \mathbb{N}$ (= set of natural numbers) by

$$q^{\sigma(x)} = \|x\| .$$

Let Γ be a Cartan subgroup of G and A its maximal split torus. Put $\overline{G} = G/A$ and let $x \mapsto \bar{x}$ denote the canonical projection of G onto $\overline{G}$. Put

$$\|\bar{x}\| = \inf_{a \in A} \|xa\| \qquad (x \in G)$$

and

$$\sigma(\bar{x}) = \inf_{a \in A} \sigma(xa) \quad .$$

Obviously $\|\bar{x}\| = q^{\sigma(\bar{x})}$ $(x \in \overline{G})$.

It is convenient to state here a few rules for $\sigma(x)$ and $\sigma(\bar{x})$. Fix $a \geq 0$. Let C_a be the set of all $x \in G$ such that $\sigma(x) \leq a$ and denote by D_a the set of all $x \in G$ such that $\sigma(\bar{x}) \leq a$. Then C_a is compact and D_a is compact mod A. Conversely, every compact subset of G is contained in some C_a for some $a \geq 0$ and every subset which is compact mod A is contained in some D_a for some $a \geq 0$.

Theorem 18. There exists a number $r \geq 0$ with the following property. Given compact subsets ω_Γ and ω of Γ and G respectively, we can choose $c \geq 0$ such that

$$|D(\gamma)|^{r/2} \|\bar{x}\| \leq c$$

if $\gamma \in \omega_\Gamma Z$, $x \in G$ and $\gamma^x \in \omega Z$.

Define $\lambda(x)$ by $q^{\lambda(x)} = |D(x)|$ $(x \in G)$.

Corollary. Given compact subsets ω_Γ and ω of Γ and G respectively, we can choose $c_o \geq 0$ such that

$$1 + \sigma(\bar{x}) \leq c_o(1 + |\lambda(\gamma)|)$$

for $x \in G$, $\gamma \in \omega_\Gamma Z$ and $\gamma^x \in \omega Z$.

This is obvious.

The proof of Theorem 18 proceeds by induction on dim G. The most important step is the proof of the following theorem.

Theorem 19. Assume prk G = 0. Suppose Γ is compact. Then we can choose c, $r \geq 0$ such that

$$|D(\gamma)|^{r/2} \|x\| \leq c \|\gamma^x\|^r$$

for $x \in G$ and $\gamma \in \Gamma$ (cf. Harish-Chandra, Am. J. Math. 79 (1957), 193-257, Lemma 5, p. 198).

Let A be any Ω-split torus in G, which is a split component of some parabolic subgroup of G and let $\mathcal{P}(A)$ denote the set of all parabolic subgroups P of G such that A is a split component of P. Let K be a compact subgroup as in the theorem of Bruhat-Tits. Let K_1 be an open subgroup of K such that $\|k\| = 1$ for all $k \in K_1$. Fix an open compact subgroup K_o of G such that

$$K_o \subset (\bar{N} \cap K_1)(M \cap K_1)(N \cap K_1)$$

for any parabolic subgroup P = MN in $\mathcal{P}(A)$.

Theorem 20. There exists a number $c \geq 1$ with the following property. Let C be a compact subset of G and y an element in G. Put

$$\sigma(C) = \sup_{x \in C} \sigma(x)$$

and $K_o(y) = K_o \cap K_o{}^y$. Let $\omega = \omega(C, y)$ denote the set of all $x \in G$ such that

$$1 + \sigma(x) \leq c(1 + \sigma(C))(1 + \sigma(y)) \quad .$$

Let f be a continuous function on G such that

(i) Supp $f \subset C\,A$ and $f(xa) = f(x)$ $(x \in G, a \in A)$.

(ii) If (P', A') is any cuspidal pair in G with $A' \subset A$, $P' \neq G$ (P' = M'N'),

then

$$\int_{N'} f(xn')dn' = 0 \qquad (x \in G) .$$

Then

$$\int_{K_o(y)} f(xky)dk = 0$$

unless $x \in \omega Z$.

This theorem is a refinement of Theorem 10 for $y = 1$. The proofs of the above results will be given in the next paragraphs.

§3. **Proof of the main theorem.**

Assuming the results of §2, we shall present first the proof of Theorem 16. Let $\omega \in {}^o\mathcal{E}(G)$ and $\pi \in \omega$ be fixed. Let Θ be its character. Put

$$\theta(x) = d(\omega)(\phi, \pi(x)\phi) \qquad (x \in G)$$

where ϕ is a K-finite unit vector in the representation space of π. Then θ is a supercusp form and we have seen (Theorem 9) that

$$\Theta(\alpha) = \int_{G/Z} dx^* \int_G \alpha(y)\theta(y^x)dy \qquad (\alpha \in C_c^\infty(G)) .$$

For any $T \geq 1$, let Ω_T denote the set of all $x \in G$ such that

$$1 + \sigma(x) \leq T .$$

Let Φ_T denote the characteristic function of the set Ω_T^*, the latter being the image of Ω_T under the canonical projection $G \longrightarrow G^* = G/Z$. We also write $\Phi_T(x) = \Phi_T(x^*)$. Then

$$\textcircled{H}(\alpha) = \lim_{T\to\infty} \int_{G/Z} \Phi_T(x^*)dx^* \int_G \alpha(y)\theta(y^x)dy$$

$$= \lim_{T\to\infty} \int_G \alpha(y)\,\textcircled{H}_T(y)dy \qquad (\alpha \in C_c^\infty(G))$$

where

$$\textcircled{H}_T(y) = \int_{G/Z} \Phi_T(x)\theta(y^x)dx^* .$$

Let Γ stand for a Cartan subgroup of G. Let A denote its maximal split torus.

Fix a compact set ω in G such that Supp $\theta \subset \omega Z$, Supp $\alpha \subset \omega$. Let ω_Γ be the set of all $\gamma \in \Gamma$ such that $\gamma^x \in \omega Z$ for some $x \in G$. Then ω_Γ is relatively compact mod Z (cf. Lemma 39). Put

$$f_\gamma(x) = \theta(\gamma^x) \qquad (\gamma \in \omega_\Gamma,\ x \in G) .$$

Now suppose $\gamma \in \Gamma$ and $y \in G$ vary in such a way that $\gamma^y \in \omega Z$. Then $\gamma \in \omega_\Gamma$ and we can choose $y_o \in yA$ such that

$$1 + \sigma(y_o) \leq c_o(1 + |\lambda(\gamma)|)$$

where c_o is a positive constant, only depending on ω and Γ (cf. the corollary of Theorem 18).

Put $\omega_\Gamma' = \omega_\Gamma \cap G'$. Then it is clear that we can choose to any $\gamma \in \omega_\Gamma'$ a compact set C_γ in G such that

(i) Supp $f_\gamma \subset C_\gamma . A$.

(ii) $1 + \sigma(C_\gamma) \leq c_o(1 + |\lambda(\gamma)|)$

(compare this with the proof of Lemma 23 where a similar result is obtained, however without any control over the size of C_γ).

Let $\Omega(\gamma)$ $(\gamma \in \omega_\Gamma')$ be the set of all $x \in G$ such that

$$1 + \sigma(x) \leq c_1(1 + |\lambda(\gamma)|)^2$$

where $c_1 = c.c_o^2$. (Here c is the constant of Theorem 20.) Let Φ_γ denote the characteristic function of $\Omega(\gamma)^* \subset G^* = G/Z$. Now suppose $\gamma \in \omega_\Gamma'$, $y \in G$

and $\gamma^y \in \omega$. Then, as we have seen above, we can choose $y_o \in yA$ such that

$$1 + \sigma(y_o) \leq c_o(1 + |\lambda(\gamma)|) .$$

Then

$$\Theta_T(\gamma^y) = \Theta_T(\gamma^{y_o}) = \int_{G/Z} \Phi_T(x)\theta(\gamma^{xy_o})dx^*$$

$$= \int_{G/Z} \Phi_T(x)dx^* \int_{K_1} \theta(\gamma^{xky_o})dk ,$$

where K_1 is chosen as in §2; its Haar measure dk is normalized such that $\int_{K_1} dk = 1$. Notice that $\sigma(xk) = \sigma(x)$ $(x \in G, k \in K_1)$. But it follows from Theorem 20 that

$$\int_{K_1} \theta(\gamma^{xky_o})dk = \int_{K_1} f_\gamma(xky_o)dk = 0$$

unless we have (mod Z):

$$1 + \sigma(xk) \leq c(1 + \sigma(C_\gamma))(1 + \sigma(y_o)) ,$$

where k runs over a set of representatives of $K_1/K_o(y_o)$ in K_1 (compare this again with the proof of Lemma 23).

Now $\sigma(xk) = \sigma(x)$ and

$$c(1 + \sigma(C_\gamma))(1 + \sigma(y_o)) \leq c.\, c_o^2(1 + |\lambda(\gamma)|)^2 .$$

Therefore $\int_{K_1} \theta(\gamma^{xky_o}) = 0$ unless $x^* \in \Omega(\gamma)^*$, i.e. unless $\Phi_\gamma(x) \neq 0$. Hence

$$\Theta_T(\gamma^y) = \int_{G/Z} \Phi_T(x)\Phi_\gamma(x)dx^* \int_{K_1} \theta(\gamma^{xky_o})dk$$

$$= \int_{G/Z} \Phi_T(x)\Phi_\gamma(x)\theta(\gamma^{xy_o})dx^*$$

$$= \int_{G/Z} \Phi_T(xy_o^{-1})\Phi_\gamma(xy_o^{-1})\theta(\gamma^x)dx^* \qquad (1)$$

As a first result, we obtain for γ and y fixed ($\gamma \in \Gamma'$, $y \in G$, $\gamma^y \in \omega$)

$$\Theta_T(\gamma^y) = \int_{G/Z} \Phi_\gamma(x)dx^* \int_{K_1} \theta(\gamma^{xky_o})dk = \int_{G/Z} dx^* \int_{K_1} \theta((\gamma^{y_o})^{xk})dk$$

as soon as $T \geq c_1(1 + |\lambda(\gamma)|)^2$. Therefore $\lim_{T\to\infty} \Theta_T(\gamma^y)$ exists. It is clear that

$$\lim_{T\to\infty} \Theta_T(\gamma^y) = F(\gamma^y)$$

(cf. Part V, 1.c.). Hence

$$\lim_{T\to\infty} \Theta_T(x) = F(x) \quad \text{for all } x \in \omega' = \omega \cap G' \;;$$

hence (by enlarging ω)

$$\lim_{T\to\infty} \Theta_T(x) = F(x) \quad \text{for all } x \in G' . \tag{2}$$

We will have an estimation for Θ_T to be able to apply Lebesgue's Theorem on dominated convergence.

From (1) we get

$$|\Theta_T(\gamma^y)| \leq \int_{G/Z} \Phi_\gamma(xy_o^{-1})|\theta(\gamma^x)|dx^*$$

$$= \int_{G/A} |\theta(\gamma^x)|d\bar{x} \int_{A/Z} \Phi_\gamma(xay_o^{-1})da^* .$$

We assume here the invariant measures $d\bar{x}$, dx^* and da^* on G/A, G/Z and A/Z respectively to be normalized such that $dx^* = d\bar{x}da^*$. Now $\theta(\gamma^x) = f_\gamma(x) = 0$ unless $x \in C_\gamma.A$. Suppose $x \in C_\gamma$. Then $\Phi_\gamma(xay_o^{-1}) = 0$ unless $(xay_o^{-1})^* \in \Omega_\gamma^*$. Put $\sigma(a^*) = \inf_{z\in Z} \|az\|$ $(a \in A)$. Since

$$1 + \sigma(a^*) \leq [1 + \sigma(x)][1 + \sigma((xay_o^{-1})^*)][1 + \sigma(y_o)] ,$$

we conclude that $\Phi_\gamma(xay_o^{-1}) = 0$ unless

$$1 + \sigma(a^*) \leq c_2(1 + |\lambda(\gamma)|)^4 \quad \text{where} \quad c_2 = c_1 c_o^2 .$$

Therefore

$$\int_{A/Z} \Phi_\gamma(xay_o^{-1})da^* \leq \int_{1+\sigma(a^*) \leq c_2(1+|\lambda(\gamma)|)^4} da^* \leq c_3(1 + |\lambda(\gamma)|)^{4\ell} .$$

where c_3 is a positive constant, not depending on the choice of $\gamma \in \omega_\Gamma'$, and $\ell = \dim A/Z$. This proves that

$$|\Theta_T(\gamma^y)| \leq c_3(1 + |\lambda(\gamma)|)^{4\ell} \int_{G/A} |\theta(\gamma^x)| d\bar{x} \qquad (\gamma \in \omega_\Gamma') .$$

By Theorem 14 we have:

$$\sup_{\gamma \in \omega_\Gamma'} |D(\gamma)|^{1/2} \int_{G/A} |\theta(\gamma^x)| d\bar{x} < +\infty .$$

Hence

$$|\Theta_T(\gamma^y)| \leq c_4 |D(\gamma)|^{-\frac{1}{2}} (1 + |\lambda(\gamma)|)^{4\ell}$$

for all $\gamma \in \Gamma$ and $y \in G$ such that $\gamma^y \in \omega' = \omega \cap G'$. Since there are only finitely many non-conjugate Cartan subgroups in G, there exists a constant c_5 such that

$$|\Theta_T(x)| \leq c_5 |D(x)|^{-1/2}(1 + |\lambda(x)|)^{4\ell}$$

for all $x \in \omega'$ and all $T \geq 1$.

It follows from Theorem 15 that the function

$$x \longmapsto |D(x)|^{-1/2}(1 + |\lambda(x)|)^{4\ell}$$

is locally summable on G.

By Lebesgue's Theorem we have now

(i) F is locally summable on G (by (2)),

(ii) $\Theta(\alpha) = \lim_{T\to\infty} \int_G \alpha(y)\, \Theta_T(y)dy = \int_G \alpha(y)F(y)dy.$

This completes the proof of the theorem.

§4. Proof of Lemma 46.

We keep the notations of §2.

The following lemma is needed in the proof of Theorem 18. Let $P = MN$ be a parabolic subgroup of G with split component A. Let $\mathfrak{n}$ be the Lie algebra of N.

Lemma 46. *There exists a number $r \geq 0$ with the following property. Given a compact subset ω_M of M, we can choose $c \geq 0$ such that*

$$|\det(\mathrm{Ad}(m^{-1}) - 1)_{\mathfrak{n}}|^r \|n\| \leq c \|m^{-1}m^n\|^r$$

for $m \in \omega_M . Z$ *and* $n \in N$.

Since $m^{-1}m^n = m^{-1}nmn^{-1}$, it is obviously enough to prove the above inequality for $m \in \omega_M$.

Let $\Sigma = \Sigma(P/A)$ be the set of all roots of (P, A). For each $\alpha \in \Sigma$, let the subspace $\mathfrak{n}_\alpha$ of $\mathfrak{n}$ be the root space of α. Then $\mathrm{Ad}(m^{-1})$ $(m \in M)$ leaves $\mathfrak{n}_\alpha$ invariant. Choose a base $(X_i)_{i \in J}$ for $\mathfrak{n}$ over Ω such that $X_i \in \mathfrak{n}_\alpha$ for some α. Let J_α be the set of all i such that $X_i \in \mathfrak{n}_\alpha$. Then $J = \bigcup_{\alpha \in \Sigma} J_\alpha$ where the union is disjoint. Put

$$J_\alpha' = \bigcup_{\beta < \alpha} J_\beta \qquad (\alpha \in \Sigma) .$$

Let $\rho_{ij}(m)$ denote the matrix coefficients of $\mathrm{Ad}(m^{-1})_{\mathfrak{n}}$ w.r.t. this base so that

$$\mathrm{Ad}(m^{-1})X_j = \sum_{i \in J} X_i \rho_{ij}(m) \qquad (j \in J) .$$

Then the ρ_{ij} are regular functions on M defined over Ω. Let $(t_i)_{i \in J}$ denote the Cartesian coordinate system on $\mathfrak{n}$ w.r.t. this base.

Lemma 47. For each $\alpha \in \Sigma$, there exist rational functions $P_i (i \in J_\alpha)$ on $M \times \mathfrak{n}$, defined over Ω, with the following property.

(i) $P_i \in \Omega[\rho_{jk}, t_k]_{j, k \in J_\alpha}$.

(ii) If $X \in \mathfrak{n}$, $m \in M$ and

$$X' = \log\{\exp \mathrm{Ad}(m^{-1})X.\exp(-X)\}$$

then

$$t_i(X') = t_i((\mathrm{Ad}(m^{-1}) - 1)X) + P_i(m, X) \qquad (i \in J_\alpha) .$$

Proof. We prove this lemma by induction on α. Clearly

$$e^{\mathrm{ad}X'} = e^{\mathrm{ad}(\mathrm{Ad}(m^{-1})X)} e^{-\mathrm{ad}X} .$$

Let E_α denote the projection of $\mathfrak{n}$ on $\mathfrak{n}_\alpha$ (corresponding to $\mathfrak{n} = \sum_{\beta \in \Sigma} \mathfrak{n}_\beta$). Then, if $H \in \mathfrak{a}$ (= Lie algebra of A)

$$E_\alpha(e^{\mathrm{ad}X'} H) = E_\alpha(e^{\mathrm{ad}(\mathrm{Ad}(m^{-1})X)} e^{-\mathrm{ad}X} H) .$$

This leads to a relation

$$t_i(X') + q_i(X') = t_i((\mathrm{Ad}(m^{-1}) - 1)X) + Q_i(m, X)$$

where $q_i \in \Omega[t_j]_{j \in J_\alpha}$, and $Q_i \in \Omega[\rho_{jk}, t_k]_{j, k \in J_\alpha}$. Hence

$$t_i(X') = t_i((\mathrm{Ad}(m^{-1}) - 1)X) + Q_i(m, X) - q_i(X')$$

and taking into account that $\mathfrak{n}_\beta$ is invariant under $\mathrm{Ad}(m^{-1})$ for all $\beta < \alpha$, our assertion follows immediately by induction hypothesis.

Let $\Delta(m) = \det(\mathrm{Ad}(m^{-1}) - 1)_{\mathfrak{n}}$ $(m \in M)$ and put $\eta(m) = (\mathrm{Ad}(m^{-1}) - 1)_{\mathfrak{n}}^{-1}$ if $\Delta(m) \neq 0$.

Denote by $\eta_{ij}(m)$ the matrix coefficients of $\eta(m)$ w.r.t. the base of $\mathfrak{n}$, chosen above. Then η_{ij} $(i, j \in J_\alpha)$ is a rational function, defined over Ω, in the $\rho_{k\ell}$ $(k, \ell \in J_\alpha)$.

Corollary 1. For each $\alpha \in \Sigma$, there exist rational functions P_i' $(i \in J_\alpha)$ on $M \times \mathfrak{n}$, defined over Ω, with the following properties:

(i) $P_i' \in \Omega[\eta_{jk}, \rho_{jk}, t_k]_{j, k \in J_\alpha}'$.

(ii) $t_i((\mathrm{Ad}(m^{-1}) - 1)X) = t_i(X') + P_i'(m, X')$ $(m \in M')$.

Here M' is the set of all points $m \in M$ where $\Delta(m) \neq 0$.

This follows immediately from the above lemma by induction on α, if we take into account the invariance of $\mathfrak{n}_\beta$ under $(\mathrm{Ad}(m^{-1}) - 1)_{\mathfrak{n}}$ and its inverse $(\beta \in \Sigma,\ m \in M')$.

Corollary 2. There exist an integer $s \geq 0$ and functions

$$R_i \in \Omega[\rho_{jk}, t_k]_{j, k \in J} \qquad (i \in J)$$

such that

$$\Delta(m)^s t_i(X) = R_i(m, X') \qquad (m \in M,\ i \in J)\ .$$

This follows easily from Corollary 1, since $\eta_{ij}(m) = \dfrac{F_{ij}(m)}{\Delta(m)}$ $(m \in M')$ for some $F_{ij} \in \Omega[\rho_{k\ell}]_{k, \ell \in J}$ $(i, j \in J)$.

Now we come to the proof of Lemma 46. The statement of the lemma is a consequence of Corollary 2, if we recall that: (1) $\exp : \mathfrak{n} \longrightarrow N$ is bijective; (2) ρ_{ij} is bounded on ω_M $(i, j \in J)$; (3) $|e^X| \leq c_1(1 + |X|)^{r_1}$ and

$$|X| \leq c_2(1 + |e^X - 1|)^{r_2} \qquad (X \in \mathfrak{n})$$

where c_1, c_2, r_1, r_2 are suitable positive constants, independent of the choice of $X \in \mathfrak{n}$. (This follows from the series for exponential and logarithm.) The details are left to the reader. Here ends the proof of Lemma 46.

§5. Proof of Theorem 18 (first step).

Let us recall the statement.

Let Γ be a Cartan subgroup of G and A its maximal split torus.

Theorem 18. There exists a number $r \geq 0$ with the following property. Given compact subsets ω_{Γ} and ω of Γ and G respectively, we can choose $c \geq 0$ such that

$$|D(\gamma)|^{r/2} \|\bar{x}\| \leq c$$

if $\gamma \in \omega_{\Gamma}.Z$, $x \in G$ and $\gamma^x \in \omega.Z$.

As announced, the proof proceeds by induction on $\dim G$. Suppose $\dim G > 0$. The first step in the proof concerns the case $A \neq Z$. Let $P = MN$ be a parabolic subgroup of G with A as a split component. Then $\dim M < \dim G$ and so, by induction hypothesis, the theorem is true for (M, Γ) in place of (G, Γ).

Let K be a compact subgroup of G, chosen as in the theorem of Bruhat and Tits. Put $\omega_1 = K\omega K$ and $\omega_P = \omega_1 \cap P$. Then ω_P is a compact set. Let ω_M denote the projection of ω_P on M w.r.t. the decomposition $P = NM$. Then ω_M is also compact. By induction hypothesis there exists a number $r_1 \geq 0$ such that the theorem is true if we replace (r, G, Γ) by (r_1, M, Γ). Hence we can choose $c_1 \geq 0$ such that

$$|D_M(\gamma)|^{r_1/2} \|\bar{m}\| \leq c_1$$

if $\gamma \in \omega_{\Gamma}.A$, $m \in M$ and $\gamma^m \in \omega_M.A$.

Also (Lemma 46) we can choose $r_2 \geq 0$ and $c_2 \geq 0$ such that r_2 is independent of ω, and

$$|\det(\mathrm{Ad}(m^{-1}) - 1)_{\mathfrak{n}}|^{r_2} \|n\| \leq c_2 \|n^{-1}m^n\|^{r_2}$$

for $m \in \omega_M.Z$, $n \in N$.

Now suppose $\gamma \in \omega_\Gamma . Z$, $x \in G$ and $\gamma^x \in \omega Z$. Then since $G = K P$, $x = knm$ $(k \in K,\ n \in N,\ m \in M)$ and $\gamma^{nm} \in \omega_1 Z$. Hence $\gamma^m \in \omega_M . Z$ and

$$(\gamma^m)^{-1} \gamma^{nm} \in N \cap (\omega_M^{-1} \omega_1 Z) = \omega_N ,$$

which is a compact set. Hence we can choose $c_3 \geq 0$ such that

$$|\det(\mathrm{Ad}(\gamma^{-1}) - 1)_{\mathfrak{n}}|^{r_2} \|n\| \leq c_3$$

where γ and n vary subject to the above conditions. Also $\gamma \in \omega_\Gamma . Z \subset \omega_\Gamma . A$ and $\gamma^m \in \omega_M . Z \subset \omega_M . A$. Therefore

$$|D_M(\gamma)|^{r_1/2} \|\overline{m}\| \leq c_1 .$$

Put $r = \max(r_1, r_2)$. Since $|\det(\mathrm{Ad}(\gamma^{-1}) - 1)_{\mathfrak{n}}|$ and $|D_M(\gamma)|$ remain bounded for $\gamma \in \omega_\Gamma . Z$, it is clear that

$$|D_M(\gamma)|^{r/2} |\det(\mathrm{Ad}(\gamma^{-1}) - 1)_{\mathfrak{n}}|^r \|n\| \|\overline{m}\| \leq c_4$$

for some constant $c_4 \geq 0$.

Now, if t is an indeterminate

$$\det(t + 1 - \mathrm{Ad}(\gamma)) = \det(t + 1 - \mathrm{Ad}(\gamma))_{\mathfrak{m}} . \det(t + 1 - \mathrm{Ad}(\gamma))_{\mathfrak{g}/\mathfrak{m}} ,$$

where $\mathfrak{m}$ is the Lie algebra of M. Therefore

$$D(\gamma) = D_M(\gamma). \det((1 - \mathrm{Ad}(\gamma))(1 - \mathrm{Ad}(\gamma^{-1})))_{\mathfrak{n}} .$$

So

$$|D(\gamma)| = |D_M(\gamma)| |\det \mathrm{Ad}(\gamma)_{\mathfrak{n}}| |\det(\mathrm{Ad}(\gamma^{-1}) - 1)_{\mathfrak{n}}|^2 .$$

Since $|\det \mathrm{Ad}(\gamma)_{\mathfrak{n}}|$ remains bounded for $\gamma \in \omega_\Gamma . Z$, there exists $c_5 \geq 0$ such that

$$|D(\gamma)|^{r/2} \|n\| \|\overline{m}\| \leq c_5 .$$

Now $\|x\| \leq \|k\| \|n\| \|m\|$ and therefore

$$\|\bar{x}\| \leq \|k\| \|n\| \|\bar{m}\| .$$

Hence

$$|D(\gamma)|^{r/2} \|\bar{x}\| \leq c$$

where $c = c_5 \cdot \sup_{k \in K} \|k\|$. This proves the first step.

It remains to consider the case $A = Z$.

§6. Proof of Theorem 19.

In this paragraph we assume $\text{prk}\, G = 0$. We recall:

Theorem 19. Let B be a Cartan subgroup of G and assume that B is compact. Then we can choose $c, r \geq 0$ such that

$$|D(b)|^{r/2} \|x\| \leq c \|b^x\|^r$$

for all $x \in G$ and $b \in B$.

We need some preparation for the proof of this theorem. Let B' be the set of all points $b \in B$ where $D(b) \neq 0$.

Lemma 48. Let $P \neq G$ be a parabolic subgroup of G. Then $B' \cap P = \emptyset$.

Proof. Let A be a split component of $P = MN$. Then $\dim \underset{\sim}{A} \geq 1$. Suppose $b \in B' \cap P$. Then $b = m_o n_o$ $(m_o \in M, n_o \in N)$. Denote by $\underset{\sim}{\mathcal{Z}}(b)$ the centralizer of b in $\underset{\sim}{G}$. Then $\underset{\sim}{\mathcal{Z}}(b) \cap \underset{\sim}{N}$ is connected, hence contained in $\underset{\sim}{\mathcal{Z}}(b)^o \cap \underset{\sim}{N} = \underset{\sim}{B} \cap \underset{\sim}{N} = \{1\}$. Since b is semisimple and normalizes $\underset{\sim}{N}$, it follows that $n \mapsto b^{-1}nbn^{-1}$ is an Ω-isomorphism of algebraic varieties of N onto N (cf. Borel-Tits, 11.1). Hence we can find $n \in N$ such that $b^{-1}nbn^{-1} = n_o^{-1}$, or $b = n^{-1}m_o n$.

Replacing (b, B) by (b^n, B^n), we may assume that $b \in B' \cap M$. Then, since $\underline{B} = \mathfrak{X}(b)^o$ and $\underline{A}$ is contained in the center of $\underline{M}$, we have $\underline{A} \subset \underline{B}$. But since $\underline{B}$ is anisotropic and $\dim \underline{A} \geq 1$, this is impossible. This proves the lemma.

The next lemma is obvious, but it is convenient to state it explicitly. We keep the notations of §2.

Lemma 49. Let $\underline{U}$ and $\underline{V}$ be two (irreducible) Ω-subvarieties of $M(m)$ and $M(n)$ respectively and ρ an Ω-morphism of $\underline{U}$ into $\underline{V}$. Let U denote the set of all Ω-rational points of $\underline{U}$. Then we can choose $c, r \geq 0$ such that

$$|\rho(x)| \leq c(1 + |x|)^r$$

for all $x \in U$.

Now we start with the proof of Theorem 19.

Let K, P_o, A_o, M_o be as in the theorem of Bruhat-Tits. Fix a simple root α_o of (P_o, A_o) and let $(P, A) = (P_o, A_o)_F$ where F is the complement of α_o in $\Sigma^o = \Sigma^o(P/A)$. Then $P = MN$ is a maximal parabolic subgroup of G and there exists an absolutely irreducible, rational representation ρ of $\underline{G}$ defined over Ω on a vector space V with the following properties.

We can choose an Ω-rational base $(v_o, v_1, \ldots, v_p)$ such that:

(1) P is exactly the stabilizer of the line Ωv_o in G. Let χ denote the corresponding Ω-rational character of P so that

$$\rho(x)v_o = \chi(x)v_o \qquad (x \in P) .$$

(2) There exist Ω-rational characters χ_i $(0 \leq i \leq p)$ of A_o such that

$$\rho(a)v_i = \chi_i(a)v_i \qquad (a \in A_o) .$$

Clearly $\chi_o = \chi$ on A_o and it follows from the theory of representations that

$$\chi_i(a) = \chi(a)\xi_{\sigma_i}(a)^{-1} \qquad (1 \leq i \leq p,\ a \in A_o)$$

where

$$\sigma_i = \sum_{\alpha \in \Sigma^o} r_i(\alpha)\alpha ,$$

$r_i(\alpha)$ being integers ≥ 0 and $r_i(\alpha_o) > 0$. Let

$$\rho(x)v_o = \sum_{0 \leq i \leq p} v_i Q_i(x) \qquad (x \in G)$$

where Q_i are regular functions on G, defined over Ω. Then

$$Q_i(axa^{-1}) = \xi_{\sigma_i}(a)^{-1} Q_i(x) \qquad (a \in A_o,\ x \in G,\ 1 \leq i \leq p) .$$

Now fix $b_o \in B'$. Since

$$G = KA_o^{+}\omega_{M_o}K$$

where ω_{M_o} is a finite subset of M_o, x^{-1} can be written in the form

$$x^{-1} = k^{-1}amk_1^{-1} \qquad (x \in G)$$

where $a \in A_o^{+}$, $m \in \omega_{M_o}$ and $k, k_1 \in K$. Then

$$\| \rho(b_o^{x}) \| \geq c . \| \rho(b_o^{a^{-1}k}) \|$$

where c is a positive number independent of x. Now

$$\| \rho(b_o^{a^{-1}k}) \| \geq \max_{1 \leq i \leq p} |Q_i(b_o^{a^{-1}k})|$$

and

$$Q_i(b_o^{a^{-1}k}) = \xi_{\sigma_i}(a) Q_i(b_o^{k}) .$$

Hence

$$\|\rho(b_o^{a^{-1}k})\| \geq \inf_{1\leq i\leq p} |\xi_{\sigma_i}(a)| \max_{1\leq j\leq p} |Q_j(b_o^k)|$$

$$\geq |\xi_{a_o}(a)| \max_{1\leq j\leq p} |Q_j(b_o^k)| \ .$$

Now

$$Q_j(b_o^k) = 0 \ (1 \leq j \leq p) \ \text{ implies that } \ \rho(b_o^k)v_o \in \Omega v_o$$

and therefore $b_o^k \in P$. But $(B')^k \cap P = \emptyset$ by Lemma 48. Hence b_o^k cannot lie in P. Therefore

$$c(k) = \max_{1\leq j\leq p} |Q_j(b_o^k)| > 0 \ .$$

Obviously $c(k)$ is a continuous function of $k \in K$. Since K is compact, we conclude that

$$c_o = \inf_{k\in K} c(k) > 0 \ .$$

$$\|\rho(b_o^x)\| \geq c \ \|\rho(b_o^{a^{-1}k})\| \geq c. c_o |\xi_{a_o}(a)| \ .$$

Hence the following lemma is obvious (by Lemma 49).

Lemma 50. <u>Fix</u> $a \in \Sigma^o(P_o/A_o)$. <u>Then we can choose numbers</u> $r, c > 0$ <u>such that if</u>

$$x = k_1 a^{-1} m k_2 \qquad (k_1, k_2 \in K, \ a \in A_o^+, \ m \in \omega_{M_o})$$

<u>then</u>

$$|\xi_a(a)| \leq c. \|b_o^x\|^r \ .$$

Corollary. <u>We can choose numbers</u> $c, r \geq 0$ <u>such that</u>

$$\|x\| \leq c \, \|b_o^{\,x}\|^r$$

<u>for all</u> $x \in G$.

<u>Proof.</u> It is clear from Lemma 49 that

$$\|x\| \leq c_1 \, \|a\| \leq c_2 \prod_{\alpha \in \Sigma^o} |\xi_\alpha(a)|^{r_o}$$

where c_1, c_2, r_o are positive numbers independent of x and $a \in A_o^+$ such that $x = k_1 a^{-1} m k_2$ (as above).

The assertion is now an immediate consequence of Lemma 50.

We continue with the proof of Theorem 19. Let L be a finite field extension of Ω such that $\underset{\sim}{B}$ splits over L and let L be endowed with the unique valuation which extends that of Ω.

Obviously B_L is a split Cartan subgroup of G_L. Hence there exists a parabolic subgroup P of G_L with B_L as a split component. Then $P = B_L . N$. Then G_L/P is compact. So $G_L = \omega P = \omega N B_L$ where ω is a compact subset of G_L.

Now fix $b_o \in B'$ as before. Let $x \in G$. Then $x = knb_1$, where $k \in \omega$, $n \in N$ and $b_1 \in B_L$. Hence

$$\|b_o^{\,x}\|_L = \|b_o^{\,kn}\|_L \leq c_o \, \|n\|_L^2$$

where $c_o > 0$ is independent of x. But

$$\|b_o^{\,x}\|_L = \|b_o^{\,x}\| \;.$$

Hence

$$\|n\|_L \geq c_o^{-\frac{1}{2}} \, \|b_o^{\,x}\|^{\frac{1}{2}} \;.$$

It follows from the preceding corollary that we can choose $c_1, r_1 \geq 0$ such that

$$\|x\| \leq c_1 \|n\|_L^{r_1}$$

for all $x \in G$ and $n \in N \cap (\omega^{-1} x B_L)$.

Now apply Lemma 46 to (G_L, P_L). Then since B is a compact subset of B_L, we can choose $c_2, r_2 \geq 0$ such that

$$|\det(\mathrm{Ad}(b^{-1}) - 1)_{\mathfrak{n}}|_L^{r_2} \|n\|_L \leq c_2 \|b^n\|_L^{r_2}$$

for $b \in B$ and $n \in N$.

Now let $x \in G$ and write again $x = knb_1$ $(k \in \omega,\ n \in N,\ b_1 \in B_L)$. Then $b^x = b^{kn}$ for $b \in B$. Hence

$$\|b^n\|_L = \|b^{k^{-1}x}\|_L \leq c_3 \|b^x\|_L$$

where $c_3 = \sup_{k \in \omega} \|k\|_L^2$. Hence

$$|\det(\mathrm{Ad}(b^{-1}) - 1)_{\mathfrak{n}}|_L^{r_1 r_2} \|n\|_L^{r_1} \leq c_2^{r_1} c_3^{r_2} \|b^x\|_L^{r_1 r_2} .$$

Put $2\rho = \sum_{\alpha > 0} \alpha$ where α runs over all roots of (P, B_L). Then

$$|D(b)|_L = \prod_{\alpha>0} |1 - \xi_\alpha(b)|_L \, |1 - \xi_\alpha(b^{-1})|_L$$

$$= |\xi_{2\rho}(b)|_L \, |\prod_{\alpha>0} (\xi_\alpha(b^{-1}) - 1)|_L^2$$

$$= |\det(\mathrm{Ad}(b^{-1}) - 1)_{\mathfrak{n}}|_L^2 \qquad (b \in B) .$$

Here we make use of the fact that B is compact and therefore

$$|\xi_{2\rho}(b)|_L = 1 .$$

Now put $r = r_1 r_2$. Then

$$|D(b)|_L^{r/2} \|n\|_L^{r_1} \le c' \|b^x\|_L^r$$

where $c' = c_2^{r_1} c_3^{r_2}$. But since $\|x\| \le c_1 \|n\|_L^{r_1}$ and $|D(b)|_L = |D(b)|$, we conclude that

$$|D(b)|^{r/2} \|x\| \le c_1 c' \|b^x\|_L^r = c_1 c' \|b^x\|^r .$$

This proves Theorem 19.

§7. Proof of Theorem 18 (second step).

In this paragraph, we keep the notations of §5. We assume $A = Z$.

Lemma 51. There exist numbers c, $r \ge 0$ such that

$$|D(\gamma)|^{r/2} \|\bar{x}\| \le c \|\overline{\gamma^x}\|^r$$

for all $\gamma \in \Gamma$ and $x \in G$.

Obviously this implies the assertion of Theorem 18 in this case. Let

$${}^{o}\underset{\sim}{G} = \bigcap_{\chi \in X_\Omega(\underset{\sim}{G})} \ker \chi .$$

Then ${}^{o}\underset{\sim}{G}$ is a connected, reductive Ω-group and $\underset{\sim}{G} = {}^{o}\underset{\sim}{G}.\underset{\sim}{A}$. Put ${}^{o}G = G \cap {}^{o}\underset{\sim}{G}$. Then ${}^{o}G.A$ is a subgroup of finite index in G.

Lemma 52. We can choose c_1, $r_1 \ge 0$ such that

$$\|x\| \le c_1 \|\bar{x}\|^{r_1}$$

for $x \in {}^{o}G$.

Proof. We may evidently assume that A is diagonal. Let n be such that $G \subset GL(n, \Omega)$. Put $V = \Omega^n$.

There exist distinct elements $\xi_i \in X_\Omega(\underset{\sim}{A})$ $(1 \le i \le p)$ and subspaces

V_i $(1 \leq i \leq p)$ of V such that

$$a.v = \xi_i(a)v \qquad (a \in A,\ v \in V_i)$$

and

$$V = \sum_{1 \leq i \leq p} V_i ,$$

where the sum is direct. Since $A = Z$, V_i is stable under G for all i. Let ρ_i denote the corresponding representation of G on V_i. Then ρ_i is defined over Ω. Hence $\det \rho_i(x) = 1$ for $x \in {}^0G$.

Choose a base for V composed of bases for V_i $(1 \leq i \leq p)$. Then w.r.t. this base

$$|x| = \sup_{1 \leq i \leq p} |\rho_i(x)| \qquad (x \in G) .$$

Now if $x \in {}^0G$, then $\det \rho_i(x) = 1$ and therefore $|\rho_i(x)| \geq 1$. Hence we can choose c_1, $r_1 \geq 1$ such that

$$|\rho_i(x^{-1})| \leq c_1 |\rho_i(x)|^{r_1} \qquad (x \in {}^0G,\ 1 \leq i \leq p) .$$

Now let $x \in {}^0G$ and $a \in A$. Then $\rho_i(xa) = \rho_i(x)\xi_i(a)$. Hence

$$\|\rho_i(xa)\| \geq \min(|\rho_i(x)|,\ |\rho_i(x^{-1})|) = |\rho_i(y_i)|$$

where y_i is either x or x^{-1}. Hence

$$c_1 \|\rho_i(xa)\|^{r_1} \geq c_1 |\rho_i(y_i)|^{r_1} \geq \|\rho_i(x)\| .$$

This shows that

$$\|x\| \leq c_1 \|xa\|^{r_1}$$

for $x \in {}^0G$ and $a \in A$ and so our assertion follows.

Now we come to the proof of Lemma 51. Let y_i $(1 \leq i \leq s)$ be a complete set of representatives for $G/{}^oG.A$. Put ${}^o\Gamma = \Gamma \cap {}^oG$. Then ${}^o\Gamma$ is a compact Cartan subgroup of ${}^oG$ and $\Gamma = {}^o\Gamma.A$. Let $x \in G$ and $\gamma \in \Gamma$. Then $x = y_i x_o a$, $\gamma = \gamma_o a_o$ where $x_o \in {}^oG$, $\gamma_o \in {}^o\Gamma$ and $a, a_o \in A$. Then

$$\|\bar{x}\| \leq \|y_i\| \, \|x_o\| \leq c_o \, \|x_o\| \; ,$$

where $c_o = \max_{1 \leq i \leq s} \|y_i\|$.

Moreover

$$\|\gamma^{x_o} a\| = \|(\gamma^x)^{y_i^{-1}} a\| \leq \|y_i\|^2 \|\gamma^x a\| \; ,$$

hence

$$\|\overline{\gamma^{x_o}}\| \leq c_o^2 \, \|\overline{\gamma^x}\| \, .$$

Therefore, by Lemma 52,

$$c_1 c_o^{2r} \|\overline{\gamma^x}\|^{r_1} \geq c_1 \, \|\overline{\gamma^{x_o}}\|^{r_1} \geq \|\gamma_o^{x_o}\| \; .$$

Now, applying Theorem 19 to $({}^oG, {}^o\Gamma)$ we get

$$c c_1 c_o^{2r+1} \|\overline{\gamma^x}\|^{rr_1} \geq c_o \, |D_o(\gamma_o)|^{r/2} \, \|x_o\| \geq |D(\gamma)|^{r/2} \, \|\bar{x}\| \; .$$

Here D_o is the "D"-function of ${}^oG$.

Since Γ/A is compact, $|D(\gamma)|$ remains bounded for $\gamma \in \Gamma$. Hence there exist constants c', $r' \geq 0$ such that

$$|D(\gamma)|^{r'/2} \, \|\bar{x}\| \leq c' \, |\overline{\gamma^x}|^{r'}$$

for all $\gamma \in \Gamma$ and $x \in G$.

This completes the proof of Lemma 51 and Theorem 18.

§8. Proof of Theorem 20.

First we recall the statement to be proved.

Let A be a split torus in G, which is a split component of some parabolic subgroup of G.

Denote by $\mathcal{P} = \mathcal{P}(A)$ the set of all parabolic subgroups P of G which have A as a split component and let $\mathcal{P}' = \mathcal{P}'(A)$ be the set of all cuspidal pairs (P', A') such that $A' \subset A$, $P' \neq G$. Let K be a compact subgroup as given by Bruhat-Tits. Let K_1 be an open subgroup of K such that $\|k\| = 1$ for all $k \in K_1$. Fix an open compact subgroup K_o of G such that

$$K_o \subset (\overline{N} \cap K_1)(M \cap K_1)(N \cap K_1)$$

for all $P \in \mathcal{P}'$, $P = MN$. (Notice that $\mathcal{P}'$ is a finite set.) For any compact set $C \subset G$, denote by Φ_C the set of all continuous functions f on G satisfying the following conditions:

(i) $f(xa) = f(x)$ $(a \in A,\ x \in G)$, $\operatorname{Supp} f \subset CA$.

(ii) If $P' \in \mathcal{P}'$ and $P' = M'N'$, then

$$\int_{N'} f(xn')dn' = 0 \qquad (x \in G) .$$

Theorem 20. <u>There exists a number</u> $c \geq 1$ <u>with the following property. Let</u> C <u>be a compact subset of</u> G <u>and</u> $y \in G$. <u>Put</u>

$$\sigma(C) = \sup_{x \in C} \sigma(x)$$

<u>and</u> $K_o(y) = K_o \cap K_o^{\,y}$. <u>Let</u> $\omega = \omega(C, y)$ <u>denote the set of all</u> $x \in G$ <u>such that</u>

$$1 + \sigma(x) \leq c.(1 + \sigma(C))(1 + \sigma(y)) .$$

<u>Then for any</u> $f \in \Phi_C$,

$$\int_{K_o(y)} f(xky)dk = 0 \qquad (x \in G)$$

<u>unless</u> $x \in \omega Z$.

<u>Proof.</u> (It may be instructive to compare the proof, given below, with the proof of Theorem 10.)

Fix $f \in \Phi_C$ and put $J(x) = \int_{K_o(y)} f(xky)dk \quad (x \in G)$. It is clear that

$$J(x) = \int_{K_o(y^{-1})} f(xyk)dk$$

where $K_o(y^{-1}) = K_o \cap K_o^{y^{-1}}$. Therefore $J(x) = 0$ unless

$$xy \in CAK_o(y^{-1}) .$$

Now $A = \bigcup_{P \in \mathcal{P}} A^+(P)$. So $xy \in \bigcup_{P \in \mathcal{P}} CA^+(P)K_o(y^{-1})$. Fix $P \in \mathcal{P}$ and put $A^+ = A^+(P)$. Let $\ell = \dim A/Z$ and let $\alpha_1, \ldots, \alpha_\ell$ be all the simple roots of (P, A). Let S be the set of all points $t = (t_1, \ldots, t_\ell)$ in $\mathbb{R}^\ell$ with $t_i \geq 0$ $(1 \leq i \leq \ell)$ and $t_1 + \ldots + t_\ell = 1$. For any $t_o \in S$ and $\varepsilon > 0$, let $S(t_o, \varepsilon)$ denote the set of all points $t \in S$ such that $|t_i - t_{oi}| < \varepsilon$ $(1 \leq i \leq \ell)$. For $a \in A^+$, put $\prod_{1 \leq i \leq \ell} |\xi_{\alpha_i}(a)| = q^{\nu(a)}$, where $\nu(a)$ is an integer ≥ 0. If $\nu(a) > 0$, let $t(a)$ denote the point in S, given by

$$|\xi_{\alpha_i}(a)| = q^{\nu(a)t_i(a)} \qquad (1 \leq i \leq \ell) .$$

Let α be a root of (P, A). Then $\alpha = \sum_{1 \leq i \leq \ell} m_i \alpha_i$ where $m_i \geq 0$ $(m_i \in \mathbb{Z})$. Put

$$\alpha(t) = \sum_{1 \leq i \leq \ell} m_i t_i = \sum_{1 \leq i \leq \ell} m_i \alpha_i(t) \qquad (t = (t_i)_{1 \leq i \leq \ell} \in S) .$$

So

$$|\xi_\alpha(a)| = \sum_{1 \leq i \leq \ell} |\xi_{\alpha_i}(a)|^{m_i} = \prod_{1 \leq i \leq \ell} q^{\nu(a)m_i t_i(a)} = q^{\nu(a)\alpha(t(a))} \qquad (a \in A^+) .$$

If $t_o \in S$ and $\varepsilon > 0$, we define $A^+(t_o, \varepsilon)$ to be the set of all $a \in A^+$ such that $t(a) \in S(t_o, \varepsilon)$ or $t(a)$ is not defined. Now fix $t_o \in S$. Then we can choose $\varepsilon > 0$ so small that

$$|\xi_\alpha(a)| \geq q^{\nu(a)\alpha(t_o)/2}$$

for $\alpha \in \Sigma(P/A)$ and $a \in A^+(t_o, \varepsilon)$.

Since S is compact and $\mathscr{P}(A)$ is a finite set, it will be enough to prove the following result.

Lemma 53. There exists a number $c \geq 1$ with the following property. Let C be a compact subset of G and $y \in G$. Let ω denote the set of all $x \in G$ such that

$$1 + \sigma(x) \leq c\,(1 + \sigma(C))(1 + \sigma(y)) .$$

Then for any $f \in \Phi_C$

$$\int_{K_o(y^{-1})} f(xyk)dk = 0$$

where $x \in CA^+(t_o, \varepsilon)K_o(y^{-1})y^{-1}$, unless $x \in \omega Z$.

We proceed with the proof of this lemma.

Without loss of generality we may assume that A is diagonal and N (= the unipotent radical of P) is upper triangular. Let F be the set of all $\alpha \in \Sigma^o = \Sigma^o(P/A)$ such that $\alpha(t_o) = 0$. Put $(P', A') = (P, A)_F$, $P' = M'N'$. Then $P' \in \mathscr{P}'$. Furthermore N' is a (normal) subgroup of N, hence triangular.

Suppose $xy \in CA^+(t_o, \varepsilon)K_o(y^{-1})$. Then $xy = \gamma a k_o$ where $\gamma \in C$, $a \in A^+(t_o, \varepsilon)$ and $k_o \in K_o(y^{-1})$. Hence if $k \in K_o(y^{-1})$,

$$xyk = \gamma a k_o k = \gamma \bar{n}_o^{\,a} m_o n_o^{\,a} a$$

where $k_o k = \bar{n}_o m_o n_o$ $(\bar{n}_o \in \bar{N}' \cap K_1,\ m_o \in M' \cap K_1,\ n_o \in N' \cap K_1)$. Put

$$\gamma_o = \gamma \bar{n}_o^{\,a} m_o .$$

Now there exists a compact set ω_o of G such that

$$(\bar{N}' \cap K)^{a_o} (M \cap K) \subset \omega_o$$

for all $a_o \in A^+$. Therefore γ_o stays in $C.\omega_o$. Define

$$J'(k) = \int_{N'\cap K_o(y^{-1})} f(xykn')dn' = \int_{N'\cap K_o(y^{-1})} f(\gamma_o(n_o n')^a)dn' \qquad (k \in K_o(y^{-1})) .$$

If $J'(k) \neq 0$, it is clear that

$$\gamma_o(n_o n')^a \in CA$$

for some $n' \in N' \cap K_o(y^{-1})$. Hence

$$(n_o n')^a \in (\gamma_o^{-1} CA) \cap N' \subset (C_1 A) \cap N'$$

where $C_1 = \omega_o^{-1} C^{-1} C$. Now put

$$\omega_{N'A} = C_1 \cap (N'A) .$$

Then $\sigma(\omega_{N'A}) \leq \sigma(C_1) \leq \sigma(\omega_o) + 2\sigma(C)$. Put

$$\omega_{N'} = (C_1 A) \cap N' .$$

Then, if $n' \in \omega_{N'}$ it is clear that $n'a \in \omega_{N'A}$ for some $a \in A$. Hence

$$\sigma(n'a) \leq \sigma(\omega_o) + 2\sigma(C) .$$

On the other hand, since A is diagonal and N is triangular, it is obvious that the diagonal entries of a and $n'a$ are the same. Hence $\|a\| \leq \|n'a\|$ and

$$\|n'\| \leq \|n'a\| \, \|a^{-1}\| \leq \|n'a\|^2 .$$

Therefore

$$\sigma(n') \leq 2\sigma(\omega_o) + 4\sigma(C) .$$

So we have

Lemma 54. $\sigma(\omega_{N'}) \leq 2\sigma(\omega_o) + 4\sigma(C)$.

We saw that $J'(k) \neq 0$ implies $(n_o n')^a \in \omega_{N'}$ for some $n' \in N' \cap K_o(y^{-1})$. So $n_o n' \in (\omega_{N'})^{a^{-1}}$ and it is clear that this set shrinks if "$a \longrightarrow \infty$". Let us find an estimation for the measure of shrinking.

Since A is diagonal and N' triangular, the diagonal of the matrix $a^{-1}n'a - 1$ $(a \in A, n \in N')$ vanishes. Furthermore

$$(a^{-1}n'a)_{ij} = a_i^{-1}a_j n_{ij}' \qquad (i \neq j) .$$

Assume $n_{ij}' \neq 0$ for some $n' \in N'$. Put $\xi_{\lambda_i}(a) = a_i$. Then $(a^{-1}n'a)_{ij} = \xi_{\lambda_i - \lambda_j}(a^{-1})n_{ij}'$ (the character group of A is written additively). Let $\Sigma(F)$ denote the set of roots of (P, A) which can be written as integer linear combination of the simple roots in F only. Let Σ' be the complement of $\Sigma(F)$ in $\Sigma = \Sigma(P/A)$. Then, since $a^{-1}\mathfrak{n}'a \subset \mathfrak{n}'$ and $\lambda_i - \lambda_j \neq 0$ $(a \in A)$, $\lambda_i - \lambda_j = \sum_{\alpha\in\Sigma'} r_\alpha \cdot \alpha$, where r_α are integers ≥ 0, $\Sigma r_\alpha \geq 1$. Therefore

$$|\xi_{\lambda_i - \lambda_j}(a^{-1})| = \prod_{\alpha\in\Sigma'} |\xi_\alpha(a^{-1})|^{r_\alpha} \leq \sup_{\substack{\alpha\in\Sigma' \\ r_\alpha \neq 0}} |\xi_\alpha(a^{-1})| \qquad (a \in A^+) .$$

Given $\alpha \in \Sigma'$, there exists, say $\alpha_1 \in {}^cF$ such that $\alpha = r_1\alpha_1 + \ldots$ with $r_1 > 0$. Hence $|\xi_\alpha(a^{-1})| \leq |\xi_{\alpha_1}(a^{-1})|$ $(a \in A^+)$. We get

$$|(a^{-1}n'a)_{ij}| = |\xi_{\lambda_i - \lambda_j}(a^{-1})|\,|n_{ij}'| \leq |n_{ij}'| \sup_{\alpha\in{}^cF} |\xi_\alpha(a^{-1})|$$

and hence

$$|a^{-1}n'a - 1| \leq |n - 1| \sup_{\alpha\in{}^cF} |\xi_\alpha(a^{-1})| \qquad (a \in A^+, n \in N') .$$

But if $a \in A^+(t_o, \varepsilon)$,

$$|\xi_\alpha(a)| = q^{\nu(a)\alpha(t(a))} \geq q^{\nu(a)\alpha(t_o)/2} \qquad (\alpha \in {}^cF) .$$

Put

$$\delta = \frac{1}{2} \inf_{\alpha \in {}^cF} \alpha(t_o) > 0 .$$

Then

$$|a^{-1}n'a - 1| \leq q^{\sigma(n') - \nu(a)\delta} \qquad (a \in A^+(t_o, \varepsilon),\ n' \in N') .$$

We can choose a number $b \geq 0$ such that $|u - 1| \leq q^{-b}$ $(u \in G)$ implies that $u \in K_o$. Now

$$|yuy^{-1} - 1| \leq |u - 1| \, \|y\|^2 \leq q^{-b}$$

if $|u - 1| \leq q^{-b-2\sigma(y)}$. This proves that

$$|u - 1| \leq q^{-b-2\sigma(y)} \qquad (u \in G)$$

implies that $u \in K_o(y^{-1})$. Therefore the following result is obvious.

Lemma 55. <u>We can choose</u> $c_1 > 0$ <u>such that</u>

$$a^{-1}n'a \in N' \cap K_o(y^{-1})$$

<u>if</u> $a \in A^+(t_o, \varepsilon)$, $n' \in N'$, $\nu(a) \geq c_1 (\sigma(n') + \sigma(y))$, $y \in G$.

Let $x \mapsto x^*$ denote the natural projection of G on $G^* = G/Z$. Put $\|x^*\| = \inf_{z \in Z} \|xz\|$, $\sigma(x^*) = \inf_{z \in Z} \sigma(xz)$ $(x \in G)$. It is clear that we can choose $c_2 > 0$ such that

$$1 + \sigma(a^*) \leq c_2 (1 + \nu(a)) \qquad (a \in A^+) .$$

Therefore by Lemmas 54 and 55, we get the following result.

Lemma 56. <u>We can choose</u> $c_3 \geq 1$, <u>independent of</u> C <u>and</u> y, <u>such that</u>

$$1 + \sigma(a^*) \geq c_3 (1 + \sigma(C))(1 + \sigma(y)) \qquad (a \in A^+(t_o, \varepsilon))$$

implies that

$$(\omega_{N'})^{a^{-1}} \subset N' \cap K_o(y^{-1}) .$$

Now we return to the proof of Lemma 53. We wrote

$$xy = \gamma a k_o \qquad (\gamma \in C,\ a \in A^+(t_o, \epsilon),\ k_o \in K_o(y^{-1})) .$$

Then $\|x\| \leq \|xy\| \, \|y\| \leq \|\gamma\| \, \|a\| \, \|y\|$ since $k_o \in K_1$ and therefore $\|k_o\| = 1$. So

$$\sigma(x) \leq \sigma(C) + \sigma(a) + \sigma(y) .$$

Hence

$$\sigma(x^*) \leq \sigma(C) + \sigma(a^*) + \sigma(y) ,$$

or

$$\sigma(a^*) \geq \sigma(x^*) - \sigma(C) - \sigma(y)$$

So, if

$$1 + \sigma(x^*) \geq 2c_3(1 + \sigma(C))(1 + \sigma(y))$$

it follows that

$$1 + \sigma(a^*) \geq c_3(1 + \sigma(C))(1 + \sigma(y))$$

and therefore by Lemma 56

$$(\omega_{N'})^{a^{-1}} \subset N' \cap K_o(y^{-1}) .$$

Lemma 57. *Suppose* x, y *are two elements in* G *such that* $xy \in CA^+(t_o, \epsilon)K_o(y^{-1})$ *and*

$$1 + \sigma(x^*) \geq 2c_3(1 + \sigma(C))(1 + \sigma(y)) .$$

Then

$$\int_{N'\cap K_o(y^{-1})} f(xykn')dn' = 0$$

for all $k \in K_o(y^{-1})$.

Proof. Fix $k \in K_o(y^{-1})$ and assume

$$0 \neq J'(k) = \int_{N'\cap K_o(y^{-1})} f(xykn')dn' \ .$$

Then, as before,

$$J'(k) = \int_{N'\cap K_o(y^{-1})} f(\gamma_o(n_o n')^a)dn' \ ,$$

where $\gamma_o \in C.\omega_o$. Since $f \in \Phi_C$, we have

$$\int_{N'} f(\gamma_o(n_o n')^a)dn' = 0 \ .$$

Furthermore

$$f(\gamma_o(n_o n')^a) = 0 \text{ unless } \gamma_o(n_o n')^a \in CA \ ,$$

i.e., unless $n_o n' \in (\omega_{N'})^{a^{-1}}$. But we have seen that

$$1 + \sigma(x^*) \geq 2c_3(1 + \sigma(C))(1 + \sigma(y))$$

implies that $(\omega_{N'})^{a^{-1}} \subset N' \cap K_o(y^{-1})$. This proves that $f(\gamma_o(n_o n')^a) = 0$ for $n' \in N'$ unless

$$n_o n' \in N' \cap K_o(y^{-1}) \ .$$

So $J'(k) = 0$ unless $n_o \in N' \cap K_o(y^{-1})$.

Now suppose $n_o \in N' \cap K_o(y^{-1})$. Then

$$f(\gamma_o(n_o n')^a) = 0 \qquad (n' \in N')$$

unless $n' \in N' \cap K_o(y^{-1})$. Hence

$$J'(k) = \int_{N' \cap K_o(y^{-1})} f(\gamma_o(n_o n')^a)dn' = \int_{N'} f(\gamma_o(n_o n')^a)dn' = 0 .$$

This contradicts our assumption.

To complete the proof of Lemma 53, write

$$J(x) = \int_{K_o(y^{-1})} f(xyk)dk = \int_{K_o(y^{-1})/K_o(y^{-1})\cap N'} d\dot{k} \int_{K_o(y^{-1})\cap N'} f(xykn')dn' .$$

Lemma 53 follows now easily from Lemma 57 if we take $c = 2c_3$. This completes the proof of Theorem 20.

Lecture Notes in Mathematics

Bisher erschienen/Already published

Vol. 1: J. Wermer, Seminar über Funktionen-Algebren. IV, 30 Seiten. 1964. DM 3,80 / $ 1.10

Vol. 2: A. Borel, Cohomologie des espaces localement compacts d'après. J. Leray. IV, 93 pages. 1964. DM 9,– / $ 2.60

Vol. 3: J. F. Adams, Stable Homotopy Theory. Third edition. IV, 78 pages. 1969. DM 8,– / $ 2.20

Vol. 4: M. Arkowitz and C. R. Curjel, Groups of Homotopy Classes. 2nd. revised edition. IV, 36 pages. 1967. DM 4,80 / $ 1.40

Vol. 5: J.-P. Serre, Cohomologie Galoisienne. Troisième édition. VIII, 214 pages. 1965. DM 18,– / $ 5.00

Vol. 6: H. Hermes, Term Logic with Choise Operator. III, 55 pages. 1970. DM 6,– / $ 1.70

Vol. 7: Ph. Tondeur, Introduction to Lie Groups and Transformation Groups. Second edition. VIII, 176 pages. 1969. DM 14,– / $ 3.80

Vol. 8: G. Fichera, Linear Elliptic Differential Systems and Eigenvalue Problems. IV, 176 pages. 1965. DM 13,50 / $ 3.80

Vol. 9: P. L. Ivănescu, Pseudo-Boolean Programming and Applications. IV, 50 pages. 1965. DM 4,80 / $ 1.40

Vol. 10: H. Lüneburg, Die Suzukigruppen und ihre Geometrien. VI, 111 Seiten. 1965. DM 8,– / $ 2.20

Vol. 11: J.-P. Serre, Algèbre Locale. Multiplicités. Rédigé par P. Gabriel. Seconde édition. VIII, 192 pages. 1965. DM 12,– / $ 3.30

Vol. 12: A. Dold, Halbexakte Homotopiefunktoren. II, 157 Seiten. 1966. DM 12,– / $ 3.30

Vol. 13: E. Thomas, Seminar on Fiber Spaces. IV, 45 pages. 1966. DM 4,80 / $ 1.40

Vol. 14: H. Werner, Vorlesung über Approximationstheorie. IV, 184 Seiten und 12 Seiten Anhang. 1966. DM 14,– / $ 3.90

Vol. 15: F. Oort, Commutative Group Schemes. VI, 133 pages. 1966. DM 9,80 / $ 2.70

Vol. 16: J. Pfanzagl and W. Pierlo, Compact Systems of Sets. IV, 48 pages. 1966. DM 5,80 / $ 1.60

Vol. 17: C. Müller, Spherical Harmonics. IV, 46 pages. 1966. DM 5,– / $ 1.40

Vol. 18: H.-B. Brinkmann und D. Puppe, Kategorien und Funktoren. XII, 107 Seiten. 1966. DM 8,– / $ 2.20

Vol. 19: G. Stolzenberg, Volumes, Limits and Extensions of Analytic Varieties. IV, 45 pages. 1966. DM 5,40 / $ 1.50

Vol. 20: R. Hartshorne, Residues and Duality. VIII, 423 pages. 1966. DM 20,– / $ 5.50

Vol. 21: Seminar on Complex Multiplication. By A. Borel, S. Chowla, C. S. Herz, K. Iwasawa, J.-P. Serre. IV, 102 pages. 1966. DM 8,– / $ 2.20

Vol. 22: H. Bauer, Harmonische Räume und ihre Potentialtheorie. IV, 175 Seiten. 1966. DM 14,– / $ 3.90

Vol. 23: P. L. Ivănescu and S. Rudeanu, Pseudo-Boolean Methods for Bivalent Programming. 120 pages. 1966. DM 10,– / $ 2.80

Vol. 24: J. Lambek, Completions of Categories. IV, 69 pages. 1966. DM 6,80 / $ 1.90

Vol. 25: R. Narasimhan, Introduction to the Theory of Analytic Spaces. IV, 143 pages. 1966. DM 10,– / $ 2.80

Vol. 26: P.-A. Meyer, Processus de Markov. IV, 190 pages. 1967. DM 15,– / $ 4.20

Vol. 27: H. P. Künzi und S. T. Tan, Lineare Optimierung großer Systeme. VI, 121 Seiten. 1966. DM 12,– / $ 3.30

Vol. 28: P. E. Conner and E. E. Floyd, The Relation of Cobordism to K-Theories. VIII, 112 pages. 1966. DM 9,80 / $ 2.70

Vol. 29: K. Chandrasekharan, Einführung in die Analytische Zahlentheorie. VI, 199 Seiten. 1966. DM 16,80 / $ 4.70

Vol. 30: A. Frölicher and W. Bucher, Calculus in Vector Spaces without Norm. X, 146 pages. 1966. DM 12,– / $ 3.30

Vol. 31: Symposium on Probability Methods in Analysis. Chairman. D. A. Kappos.IV, 329 pages. 1967. DM 20,– / $ 5.50

Vol. 32: M. André, Méthode Simpliciale en Algèbre Homologique et Algèbre Commutative. IV, 122 pages. 1967. DM 12,– / $ 3.30

Vol. 33: G. I. Targonski, Seminar on Functional Operators and Equations. IV, 110 pages. 1967. DM 10,– / $ 2.80

Vol. 34: G. E. Bredon, Equivariant Cohomology Theories. VI, 64 pages. 1967. DM 6,80 / $ 1.90

Vol. 35: N. P. Bhatia and G. P. Szegö, Dynamical Systems. Stability Theory and Applications. VI, 416 pages. 1967. DM 24,– / $ 6.60

Vol. 36: A. Borel, Topics in the Homology Theory of Fibre Bundles. VI, 95 pages. 1967. DM 9,– / $ 2.50

Vol. 37: R. B. Jensen, Modelle der Mengenlehre. X, 176 Seiten. 1967. DM 14,– / $ 3.90

Vol. 38: R. Berger, R. Kiehl, E. Kunz und H.-J. Nastold, Differentialrechnung in der analytischen Geometrie IV, 134 Seiten. 1967 DM 12,– / $ 3.30

Vol. 39: Séminaire de Probabilités I. II, 189 pages. 1967. DM 14,– / $ 3.90

Vol. 40: J. Tits, Tabellen zu den einfachen Lie Gruppen und ihren Darstellungen. VI, 53 Seiten. 1967. DM 6,80 / $ 1.90

Vol. 41: A. Grothendieck, Local Cohomology. VI, 106 pages. 1967. DM 10,– / $ 2.80

Vol. 42: J. F. Berglund and K. H. Hofmann, Compact Semitopological Semigroups and Weakly Almost Periodic Functions. VI, 160 pages. 1967. DM 12,– / $ 3.30

Vol. 43: D. G. Quillen, Homotopical Algebra. VI, 157 pages. 1967. DM 14,– / $ 3.90

Vol. 44: K. Urbanik, Lectures on Prediction Theory. IV, 50 pages. 1967. DM 5,80 / $ 1.60

Vol. 45: A. Wilansky, Topics in Functional Analysis. VI, 102 pages. 1967. DM 9,60 / $ 2.70

Vol. 46: P. E. Conner, Seminar on Periodic Maps. IV, 116 pages. 1967. DM 10,60 / $ 3.00

Vol. 47: Reports of the Midwest Category Seminar I. IV, 181 pages. 1967. DM 14,80 / $ 4.10

Vol. 48: G. de Rham, S. Maumary et M. A. Kervaire, Torsion et Type Simple d'Homotopie. IV, 101 pages. 1967. DM 9,60 / $ 2.70

Vol. 49: C. Faith, Lectures on Injective Modules and Quotient Rings. XVI, 140 pages. 1967. DM 12,80 / $ 3.60

Vol. 50: L. Zalcman, Analytic Capacity and Rational Approximation. VI, 155 pages. 1968. DM 13,20 / $ 3.70

Vol. 51: Séminaire de Probabilités II. IV, 199 pages. 1968. DM 14,– / $ 3.90

Vol. 52: D. J. Simms, Lie Groups and Quantum Mechanics. IV, 90 pages. 1968. DM 8,– / $ 2.20

Vol. 53: J. Cerf, Sur les difféomorphismes de la sphère de dimension trois ($\Gamma_4 = 0$). XII, 133 pages. 1968. DM 12,– / $ 3.30

Vol. 54: G. Shimura, Automorphic Functions and Number Theory. VI, 69 pages. 1968. DM 8,– / $ 2.20

Vol. 55: D. Gromoll, W. Klingenberg und W. Meyer, Riemannsche Geometrie im Großen. VI, 287 Seiten. 1968. DM 20,– / $ 5.50

Vol. 56: K. Floret und J. Wloka, Einführung in die Theorie der lokalkonvexen Räume. VIII, 194 Seiten. 1968. DM 16,– / $ 4.40

Vol. 57: F. Hirzebruch und K. H. Mayer, O (n)-Mannigfaltigkeiten, exotische Sphären und Singularitäten. IV, 132 Seiten. 1968. DM 10,80 / $ 3.00

Vol. 58: Kuramochi Boundaries of Riemann Surfaces. IV, 102 pages. 1968. DM 9,60 / $ 2.70

Vol. 59: K. Jänich, Differenzierbare G-Mannigfaltigkeiten. VI, 89 Seiten. 1968. DM 8,– / $ 2.20

Vol. 60: Seminar on Differential Equations and Dynamical Systems. Edited by G. S. Jones. VI, 106 pages. 1968. DM 9,60 / $ 2.70

Vol. 61: Reports of the Midwest Category Seminar II. IV, 91 pages. 1968. DM 9,60 / $ 2.70

Vol. 62: Harish-Chandra, Automorphic Forms on Semisimple Lie Groups X, 138 pages. 1968. DM 14,– / $ 3.90

Vol. 63: F. Albrecht, Topics in Control Theory. IV, 65 pages. 1968. DM 6,80 / $ 1.90

Vol. 64: H. Berens, Interpolationsmethoden zur Behandlung von Approximationsprozessen auf Banachräumen. VI, 90 Seiten. 1968. DM 8,– / $ 2.20

Vol. 65: D. Kölzow, Differentiation von Maßen. XII, 102 Seiten. 1968. DM 8,– / $ 2.20

Vol. 66: D. Ferus, Totale Absolutkrümmung in Differentialgeometrie und -topologie. VI, 85 Seiten. 1968. DM 8,– / $ 2.20

Vol. 67: F. Kamber and P. Tondeur, Flat Manifolds. IV, 53 pages. 1968. DM 5,80 / $ 1.60

Vol. 68: N. Boboc et P. Mustaţă, Espaces harmoniques associés aux opérateurs différentiels linéaires du second ordre de type elliptique. VI, 95 pages. 1968. DM 8,60 / $ 2.40

Vol. 69: Seminar über Potentialtheorie. Herausgegeben von H. Bauer. VI, 180 Seiten. 1968. DM 14,80 / $ 4.10

Vol. 70: Proceedings of the Summer School in Logic. Edited by M. H. Löb. IV, 331 pages. 1968. DM 20,– / $ 5.50

Vol. 71: Séminaire Pierre Lelong (Analyse), Année 1967 – 1968. VI, 19 pages. 1968. DM 14,– / $ 3.90

Bitte wenden / Continued

Vol. 72: The Syntax and Semantics of Infinitary Languages. Edited by J. Barwise. IV, 268 pages. 1968. DM 18,– / $ 5.00

Vol. 73: P. E. Conner, Lectures on the Action of a Finite Group. IV, 123 pages. 1968. DM 10,– / $ 2.80

Vol. 74: A. Fröhlich, Formal Groups. IV, 140 pages. 1968. DM 12,– / $ 3.30

Vol. 75: G. Lumer, Algèbres de fonctions et espaces de Hardy. VI, 80 pages. 1968. DM 8,– / $ 2.20

Vol. 76: R. G. Swan, Algebraic K-Theory. IV, 262 pages. 1968. DM 18,– / $ 5.00

Vol. 77: P.-A. Meyer, Processus de Markov: la frontière de Martin. IV, 123 pages. 1968. DM 10,– / $ 2.80

Vol. 78: H. Herrlich, Topologische Reflexionen und Coreflexionen. XVI, 166 Seiten. 1968. DM 12,– / $ 3.30

Vol. 79: A. Grothendieck, Catégories Cofibrées Additives et Complexe Cotangent Relatif. IV, 167 pages. 1968. DM 12,– / $ 3.30

Vol. 80: Seminar on Triples and Categorical Homology Theory. Edited by B. Eckmann. IV, 398 pages. 1969. DM 20,– / $ 5.50

Vol. 81: J.-P. Eckmann et M. Guenin, Méthodes Algébriques en Mécanique Statistique. VI, 131 pages. 1969. DM 12,– / $ 3.30

Vol. 82: J. Wloka, Grundräume und verallgemeinerte Funktionen. VIII, 131 Seiten. 1969. DM 12,– / $ 3.30

Vol. 83: O. Zariski, An Introduction to the Theory of Algebraic Surfaces. IV, 100 pages. 1969. DM 8,– / $ 2.20

Vol. 84: H. Lüneburg, Transitive Erweiterungen endlicher Permutationsgruppen. IV, 119 Seiten. 1969. DM 10,– / $ 2.80

Vol. 85: P. Cartier et D. Foata, Problèmes combinatoires de commutation et réarrangements. IV, 88 pages. 1969. DM 8,– / $ 2.20

Vol. 86: Category Theory, Homology Theory and their Applications I. Edited by P. Hilton. VI, 216 pages. 1969. DM 16,– / $ 4.40

Vol. 87: M. Tierney, Categorical Constructions in Stable Homotopy Theory. IV, 65 pages. 1969. DM 6,– / $ 1.70

Vol. 88: Séminaire de Probabilités III. IV, 229 pages. 1969. DM 18,– / $ 5.00

Vol. 89: Probability and Information Theory. Edited by M. Behara, K. Krickeberg and J. Wolfowitz. IV, 256 pages. 1969. DM 18,– / $ 5.00

Vol. 90: N. P. Bhatia and O. Hajek, Local Semi-Dynamical Systems. II, 157 pages. 1969. DM 14,– / $ 3.90

Vol. 91: N. N. Janenko, Die Zwischenschrittmethode zur Lösung mehrdimensionaler Probleme der mathematischen Physik. VIII, 194 Seiten. 1969. DM 16,80 / $ 4.70

Vol. 92: Category Theory, Homology Theory and their Applications II. Edited by P. Hilton. V, 308 pages. 1969. DM 20,– / $ 5.50

Vol. 93: K. R. Parthasarathy, Multipliers on Locally Compact Groups. III, 54 pages. 1969. DM 5,60 / $ 1.60

Vol. 94: M. Machover and J. Hirschfeld, Lectures on Non-Standard Analysis. VI, 79 pages. 1969. DM 6,– / $ 1.70

Vol. 95: A. S. Troelstra, Principles of Intuitionism. II, 111 pages. 1969. DM 10,– / $ 2.80

Vol. 96: H.-B. Brinkmann und D. Puppe, Abelsche und exakte Kategorien, Korrespondenzen. V, 141 Seiten. 1969. DM 10,– / $ 2.80

Vol. 97: S. O. Chase and M. E. Sweedler, Hopf Algebras and Galois theory. II, 133 pages. 1969. DM 10,– / $ 2.80

Vol. 98: M. Heins, Hardy Classes on Riemann Surfaces. III, 106 pages. 1969. DM 10,– / $ 2.80

Vol. 99: Category Theory, Homology Theory and their Applications III. Edited by P. Hilton. IV, 489 pages. 1969. DM 24,– / $ 6.60

Vol. 100: M. Artin and B. Mazur, Etale Homotopy. II, 196 Seiten. 1969. DM 12,– / $ 3.30

Vol. 101: G. P. Szegö et G. Treccani, Semigruppi di Trasformazioni Multivoche. VI, 177 pages. 1969. DM 14,– / $ 3.90

Vol. 102: F. Stummel, Rand- und Eigenwertaufgaben in Sobolewschen Räumen. VIII, 386 Seiten. 1969. DM 20,– / $ 5.50

Vol. 103: Lectures in Modern Analysis and Applications I. Edited by C. T. Taam. VII, 162 pages. 1969. DM 12,– / $ 3.30

Vol. 104: G. H. Pimbley, Jr., Eigenfunction Branches of Nonlinear Operators and their Bifurcations. II, 128 pages. 1969. DM 10,– / $ 2.80

Vol. 105: R. Larsen, The Multiplier Problem. VII, 284 pages. 1969. DM 18,– / $ 5.00

Vol. 106: Reports of the Midwest Category Seminar III. Edited by S. Mac Lane. III, 247 pages. 1969. DM 16,– / $ 4.40

Vol. 107: A. Peyerimhoff, Lectures on Summability. III, 111 pages. 1969. DM 8,– / $ 2.20

Vol. 108: Algebraic K-Theory and its Geometric Applications. Edited by R. M. F. Moss and C. B. Thomas. IV, 86 pages. 1969. DM 6,– / $ 1.70

Vol. 109: Conference on the Numerical Solution of Differential Equations. Edited by J. Ll. Morris. VI, 275 pages. 1969. DM 18,– / $ 5.00

Vol. 110: The Many Facets of Graph Theory. Edited by G. Chartrand and S. F. Kapoor. VIII, 290 pages. 1969. DM 18,– / $ 5.00

Vol. 111: K. H. Mayer, Relationen zwischen charakteristischen Zahlen. III, 99 Seiten. 1969. DM 8,– / $ 2.20

Vol. 112: Colloquium on Methods of Optimization. Edited by N. N. Moiseev. IV, 293 pages. 1970. DM 18,– / $ 5.00

Vol. 113: R. Wille, Kongruenzklassengeometrien. III, 99 Seiten. 1970. DM 8,– / $ 2.20

Vol. 114: H. Jacquet and R. P. Langlands, Automorphic Forms on GL (2). VII, 548 pages. 1970. DM 24,– / $ 6.60

Vol. 115: K. H. Roggenkamp and V. Huber-Dyson, Lattices over Orders I. XIX, 290 pages. 1970. DM 18,– / $ 5.00

Vol. 116: Sèminaire Pierre Lelong (Analyse) Année 1969. IV, 195 pages. 1970. DM 14,– / $ 3.90

Vol. 117: Y. Meyer, Nombres de Pisot, Nombres de Salem et Analyse Harmonique. 63 pages. 1970. DM 6,– / $ 1.70

Vol. 118: Proceedings of the 15th Scandinavian Congress, Oslo 1968. Edited by K. E. Aubert and W. Ljunggren. IV, 162 pages. 1970. DM 12,– / $ 3.30

Vol. 119: M. Raynaud, Faisceaux amples sur les schémas en groupes et les espaces homogènes. III, 219 pages. 1970. DM 14,– / $ 3.90

Vol. 120: D. Siefkes, Büchi's Monadic Second Order Successor Arithmetic. XII, 130 Seiten. 1970. DM 12,– / $ 3.30

Vol. 121: H. S. Bear, Lectures on Gleason Parts. III, 47 pages. 1970. DM 6,– /$ 1.70

Vol. 122: H. Zieschang, E. Vogt und H.-D. Coldewey, Flächen und ebene diskontinuierliche Gruppen. VIII, 203 Seiten. 1970. DM 16,– / $ 4.40

Vol. 123: A. V. Jategaonkar, Left Principal Ideal Rings. VI, 145 pages. 1970. DM 12,– / $ 3.30

Vol. 124: Séminare de Probabilités IV. Edited by P. A. Meyer. IV, 282 pages. 1970. DM 20,– / $ 5.50

Vol. 125: Symposium on Automatic Demonstration. V, 310 pages. 1970. DM 20,– / $ 5.50

Vol. 126: P. Schapira, Théorie des Hyperfonctions. XI, 157 pages. 1970. DM 14,– / $ 3.90

Vol. 127: I. Stewart, Lie Algebras. IV, 97 pages. 1970. DM 10,– / $ 2.80

Vol. 128: M. Takesaki, Tomita's Theory of Modular Hilbert Algebras and its Applications. II, 123 pages. 1970. DM 10,– / $ 2.80

Vol. 129: K. H. Hofmann, The Duality of Compact Semigroups and C*- Bigebras. XII, 142 pages. 1970. DM 14,– / $ 3.90

Vol. 130: F. Lorenz, Quadratische Formen über Körpern. II, 77 Seiten. 1970. DM 8,– / $ 2.20

Vol. 131: A Borel et al., Seminar on Algebraic Groups and Related Finite Groups. VII, 321 pages. 1970. DM 22,– / $ 6.10

Vol. 132: Symposium on Optimization. III, 348 pages. 1970. DM 22,– / $ 6.10

Vol. 133: F. Topsøe, Topology and Measure. XIV, 79 pages. 1970. DM 8,– / $ 2.20

Vol. 134: L. Smith, Lectures on the Eilenberg-Moore Spectral Sequence. VII, 142 pages. 1970. DM 14,– / $ 3.90

Vol. 135: W. Stoll, Value Distribution of Holomorphic Maps into Compact Complex Manifolds. II, 267 pages. 1970. DM 18,– / $

Vol. 136: M. Karoubi et al., Séminaire Heidelberg-Saarbrücken-Strasbuorg sur la K-Théorie. IV, 264 pages. 1970. DM 18,– / $ 5.00

Vol. 137: Reports of the Midwest Category Seminar IV. Edited by S. MacLane. III, 139 pages. 1970. DM 12,– / $ 3.30

Vol. 138: D. Foata et M. Schützenberger, Théorie Géométrique des Polynômes Eulériens. V, 94 pages. 1970. DM 10,– / $ 2.80

Vol. 139: A. Badrikian, Séminaire sur les Fonctions Aléatoires Linéaires et les Mesures Cylindriques. VII, 221 pages. 1970. DM 18,– / $ 5.00

Vol. 140: Lectures in Modern Analysis and Applications II. Edited by C. T. Taam. VI, 119 pages. 1970. DM 10,– / $ 2.80

Vol. 141: G. Jameson, Ordered Linear Spaces. XV, 194 pages. 1970. DM 16,– / $ 4.40

Vol. 142: K. W. Roggenkamp, Lattices over Orders II. V, 388 pages. 1970. DM 22,– / $ 6.10

Vol. 143: K. W. Gruenberg, Cohomological Topics in Group Theory. XIV, 275 pages. 1970. DM 20,– / $ 5.50